Fadhil H. Easif
Saad A. Manaa
Rostam K. Saeed

Numerical Solutions and Stability Analysis of Brusselator System

Fadhil H. Easif
Saad A. Manaa
Rostam K. Saeed

Numerical Solutions and Stability Analysis of Brusselator System

Reaction-Diffusion systems

Noor Publishing

Cover image: www.ingimage.com

Publisher:
Noor Publishing
is a trademark of
Dodo Books Indian Ocean Ltd., member of the OmniScriptum S.R.L Publishing group
str. A.Russo 15, of. 61, Chisinau-2068, Republic of Moldova Europe
Printed at: see last page
ISBN: 978-620-0-77508-5

ACKNOWLEDGMENTS

Words would not be enough to express my deep feeling of gratitude and immense indebtedness to my supervisors, Professors ***Dr. Saad A. Manaa*** **and** ***Dr. Rostam K. Saeed***, for their continuous support, encouragement, many invaluable suggestions, and comments on this work and for acquainting me with the field of "***Reaction-Diffusion systems***". Without their guidance, this research could not have been carried out.

My heartfelt thanks are also due to the Dean and the staff of the College of Education/University of Duhok, especially the staff of the Department of Mathematics for their support and cooperation.

Special appreciation is to my family for their love and support over the years of my study. I could not have completed this work without their patience. My feelings toward my family cannot be expressed in words.

I would like to express my thanks to my colleagues Mr. Farhad M. Saleem Abdullah and Laya Y. Haweil for their help.

Finally, I would like to thank everyone helped me in performance of this work.

Fadhil H.Easif
2010

Dedication

To

The Spirit of My Parents.

My wife and Children.

My Brothers.

Friends.

With Love and Respect.

Fadhil

List of Abbreviations

Symbol	Description
ADE	Alternating Direction Explicit
ADI	Alternating Direction Implicit
CM	Collocation Method
FDM	Finite Difference Method
FEM	Finite Element Method
GM	Galerkin Method
LSM	Least Square Method
MSAM	Modified Successive Approximation Method.
PDE	Partial Differential Equations
PM	Partition Method
RD	Reaction Diffusion
SAM	Successive Approximation Method
WRM	Weighted Residual Method

Contents:

Chapter Four: Expansion Method

Chapter Five: Iterative Methods

Chapter Six: Numerical Examples, Conclusions and Recommendations

CHAPTER ONE
Introduction and Fundamental Concepts

1.1 Introduction

Partial Differential Equation (PDE) is a relation involving an unknown function of several independent variables and its partial derivatives with respect to those variables. Partial differential equations are used to formulate and solve problems of unknown functions of several variables such as the propagation of sound or heat, electrostatics, electrodynamics, fluid flow, elasticity, or more generally any process that is distributed in space or distributed in space and time. Very different physical problems may have an identical mathematical formulations [34].

Many physical, chemical and engineering problems can be modeled mathematically in the form of system of partial differential equations or system of ordinary differential equations. Finding the exact solution for the above problems which involve partial differential equations is difficult in some cases. Here we have to find the numerical solution of these problems using computers which came into existence [67].

For nonlinear partial differential equations, however, the linear superposition principle cannot be applied to generate a new solution. So, because most solution methods for linear equations cannot be applied to nonlinear equations, there is no general method of finding analytical solutions of nonlinear partial differential equations, and numerical techniques are usually required for their solution. A transformation of variables can, sometimes, be found, which transforms a nonlinear equation into linear equation, or some other methods can be used to find a solution of a particular nonlinear equation. In fact new methods are usually required for finding solutions of nonlinear equations [14].

Since the development of high-speed computing devices, the numerical solution of PDEs has been active with the invention of new algorithms and the

examination of the underlying theory. This area is one of the most active one in applied mathematics and it has a great impact on science and engineering because of its ease and efficiency in solving even the most complicated problems. The basic idea of the method of finite difference is to cast the continuous problem described by the PDE and auxiliary conditions into a discrete problem that can be solved by computer in finitely many steps. The discretization is accomplished by restricting the problem to a set of discrete points. By systematic procedures, we then calculate the unknown function at those discrete points. Consequently, a finite difference technique yields a solution only at discrete points in the domain of interest rather than, as we expect for an analytical calculation, a formula or closed-form solution valid at all points of the domain [46].

Methods of solutions for nonlinear equations represent only one aspect of the theory of nonlinear partial differential equations. Like linear equations, questions of existence, uniqueness, and stability of solutions of nonlinear partial differential equations are of fundamental importance. These and other aspects of nonlinear equations have led the subject into one of the most diverse and active areas of modern mathematics [14].

Various finite difference algorithms or schemes have been presented for the solution of hyperbolic-parabolic problem or its simpler derivatives, such as the classical diffusion equation. It is well-known that many of these schemes are partially unsatisfactory due to the formation of oscillations and numerical diffusion within the solution [69].

Although solution by finite difference method is more general involving stability and convergence problems, it may require special handling of boundary conditions, large computer storage and time execution. The problem of numerical dispersion for finite difference solutions is also difficult to overcome [24].

The finite element method is one of the most flexible tools available for solving engineering problems of the kind involved in analyzing the deformation of solids, the transfer of heat, the flow of fluid, electrical configuration, with boundary conditions [3].

The finite element method was developed simultaneously with the increasing use of high-speed electronic digital computers and with the growing emphasis on numerical methods for engineering analysis. Although the method was originally developed for structural analysis, the general nature of the theory on which it was based has also made possible its successful application for solutions of problems in other fields of engineering [17].

There are many problems which simply do not have analytical solutions, or those whose exact solution is beyond our current state of knowledge. There are also many problems which are too long (or tedious) to solve by hand. When such problems arise, we can exploit numerical analysis to reduce the problem to one involving a finite number of unknowns and use a computer to solve the resulting equations. Also, there are many methods for approximating the solution of the boundary value problem. Generally speaking, they fall into two classes: those in which the solution is approximated numerically at a number of discrete points called "grid, nodal, net or mesh points", and those in which the solution is approximated by a finite number of terms of infinite expansions in terms of a sequence of functions. The former are usually called the" difference methods" and the latter the "weighted residual method". The methods of weighted residuals are general techniques for developing approximate solutions of operator equations. In all of them the unknown solution is approximated by a set of local basis functions containing adjustable constants or functions. These constants or functions are chosen by various criteria to give the 'best' approximation for the selected family[32].

There is another approximation method for solving integral equations and differential equations. This method starts by using the constant function as an approximation to a solution. We substitute this approximation into the right side of the given equation and use the result as the next approximation to the solution. Then we substitute this approximation into the right side of the given equation to obtain what we hope is a still better approximation and we continue the process. Our goal is to find a function with the property that when it is substituted in the right side of the given equation, the result is the same function. This procedure is known as successive approximation method [6].

1.2 Mathematical Model

A general class of nonlinear-diffusion system is in the form:

$$\frac{\partial u}{\partial t} = d_1 \Delta u + a_1 u + b_1 v + f(u, v) + g_1(x)$$

$$\frac{\partial v}{\partial t} = d_2 \Delta u + a_2 u + b_2 v - f(u, v) + g_2(x),$$

Where Δ is laplace operator, with homogenous Dirchlet or Neumann boundary condition on a bounded domain Ω , $n \leq 3$, with locally Lipschitz continuous boundary. It is well known that reaction and diffusion of chemical or biochemical species can produce a variety of spatial patterns. This class of reaction diffusion systems includes some significant pattern formation equations arising from the modeling of kinetics of chemical or biochemical reactions and from the biological pattern formation theory.

In this group, the following four systems are typically important and serve as mathematical models in physical chemistry and in biology:

Brusselator Model

$$a_1 = -(b+1), b_1 = 0, a_2 = b, b_2 = 0, f = u^2 v, g_1 = a, g_2 = 0,$$

where *a* and *b* are positive constants.

Gray-Scott Model

$$a_1 = -(f+k), b_1 = 0, a_2 = 0, b_2 = -F, f = u^2 v, g_1 = 0, g_2 = F,$$

where *F* and *k* are positive constants.

Glycolysis Model :

$$a_1 = -1, b_1 = k, a_2 = 0, b_2 = -k, f = u^2 v, g_1 = p, g_2 = \delta,$$

Schnackenberg Model

$$a_1 = -k, b_1 = a_2 = b_2 = 0, f = u^2 v, g_1 = a, g_2 = b,$$

The Brusselator model describes the case in which the chemical reactions follow the scheme :

$$A \to U$$

$$B + U \to V + D$$

$$2U + V \to 3U$$

$$U \to E$$

where *A, B, D, E, U* and *V* are chemical components. Let $u(x,t)$ and $v(x,t)$ be the concentrations of *U* and *V*, and assume that the concentrations of the input components *A* and *B* are held constant during the reaction process, denoted by a and b respectively. Then one obtains the following system of two nonlinearly coupled reaction-diffusion equations:

$$\frac{\partial u}{\partial t} = d_1 \Delta u + u^2 v - (b+1)u + a, \quad (t,x) \in (0,\infty) \times \Omega \,,$$
$$\frac{\partial v}{\partial t} = d_2 \Delta v - u^2 v + bu, \quad (t,x) \in (0,\infty) \times \Omega, \tag{1.1}$$

Initial and boundary conditions;

$$u(t,x) = v(t,x) = 0, \;\; t > 0$$
$$u(0,x) = u_0(x), \;\; v(0,x) = v_0(x), \quad x \in \Omega, \tag{1.2}$$

and with Neumann boundary conditions;

$$\frac{\partial u}{\partial x} = 0, \quad at\, x = 0 \;\; and \;\; x = L$$
$$\frac{\partial v}{\partial x} = 0, \quad at\, x = 0 \;\; and \;\; x = L \tag{1.3}$$

where d_1, d_2, a, b and L are positive constants [80].

The system of equations (1.1) is called the Brusselator equations. There are several known examples of autocatalysis which can be modeled by the Brusselator equations, such as ferrocyanide –iodate –sulphite reactions, and fungal mycelia growth [5], [66].

A non-linear open reaction-diffusion system is known to exhibit a variety of interesting phenomena while operating far away from the thermodynamic equilibrium. These phenomena basically include symmetric and asymmetric multiple state stationary structures and temporal oscillations which consist of various type of linear and nonlinear waves, propagating fronts and pulses, and certain shock structures. Some of these phenomena have been observed experimentally and subsequently proved mathematically. A striking experimental example of the propagating waves is the Belousov-Zhabotinsky reaction [81]. While traveling concentration waves have been observed in catalytic reaction system [28]. Such solutions have been termed as "dissipative structures" by [53]. They occur as a result of the loss of stability of the system due to changes in a certain parameters.

The Brusselator chemical network is a simple tri-molecular model, originally proposed in [40], and is represented by an autocatalytic reaction mechanism. Because of it is simplicity, this model has been subjected to many theoretical investigations. By making use of the bifurcation theory, the bifurcation points and approximate steady-state solution for this model calculated analytically [5] and [52]. For the same model, [30] and [37] have numerically calculated various steady-state profiles and compared them with approximate solutions obtained analytically. A rotating wave solution has been obtained for this model [22]. A critical analysis of literature shows that in the majority of previous investigations, the diffusion coefficient of the intermediate components X and Y has been treated as constant (independent of composition) since this assumption simplifies the bifurcation analysis. However, a more realistic approach necessitates consideration of a composition –dependent diffusion coefficient. In certain real situations, the diffusion coefficient of the solute shows dependence on its concentration and hence on the spatial coordinates. To solve such nonlinear diffusion problems, [26] and [12] have suggested various numerical and approximate techniques. The dependent of the diffusion coefficients on concentration has been experimentally reported by [19], [10] and [75] for diffusion of oxygen and water in certain biological media. [45] has theoretically studied the Michaelis-Menten reaction in a spherical geometry and has considered both linear and nonlinear dependence of the diffusion coefficient on composition. His calculations indicate a decrease (increase) of concentration gradients in the system if the diffusion coefficient increase (decrease)with concentration. For the Brusselator model, [44] analysed the effect of concentration-dependent diffusion coefficient on the bifurcation pattern. Their study indicates that a variable diffusion coefficient correction may affect the stability of steady- state solutions and the occurrence of Hof bifurcation points as well [51].

Reaction-diffusion (RD) systems arise frequently in the study of chemical and biological phenomena and are naturally modeled by parabolic Partial Differential

Equations (PDEs). The dynamics of RD systems has been the subject of intense research activity over the past decades. The reason is that RD system exhibit very rich dynamic behavior including periodic and quasi-periodic solutions and chaos. The RD system whose dynamics has been studied extensively is the Brusselator reaction scheme in one and two dimensional domains. In [5] and [27] extensive bifurcation studies of the Brusselator model showed that the system exhibits a very rich dynamic behavior for different regions in the parameter space. In [9] the dynamics of the Brusselator model with Dirichlet boundary conditions was studied using the length of the domain as the bifurcation parameter and evidence of chaotic behavior was presented. Theoretical justification of the existence of a periodic solutions based on bifurcation theory was presented in [23].

Most chemical reactions can present rich phenomena in vessels such as chemical oscillations, periodic doubling, chemical waves, and chaos. Analysis of forced nonlinear oscillations plays an important role in understanding the dynamical phenomena of electronic generators, mechanical, chemical and biological systems. Even small external disturbances are likely to change behaviors of dynamical systems[71]. Even though the analysis of complex dynamics of RD systems has been a research subject for more than 30 years, the use of feedback control to suppress complex dynamics of RD systems and the study of the dynamics of RD systems under feedback control have been addressed only recently. in [57], a linear adaptive control strategy was applied to the Gray-Scott model in order to control the formation of patterns in a one-dimensional domain, while in [58], an experimental application of linear modal feedback control for suppressing chaotic temporal fluctuation of spatiotemporal thermal pattern on a catalytic wafer was reported in [9]. A feedback controller based on the singular value decomposition of the spatial differential operator was used to control complex dynamics of the Brusselator model and a detailed bifurcation analysis of the closed-loop system was performed [36].

Various orders are self-organized far from the chemical equilibrium. The theoretical procedures and notions for describing the dynamics of patterns of formation have been developed for the last three decades [53] and [13].

Attempts have also been made to understand morphological orders in biology [50]. Clarification of the mechanisms of the formation of orders and the relationship among them constitute one of the fundamental problems in non-equilibrium statistical physics.

Among a number of pattern formation phenomena ranging from crystal growth to hydrodynamic systems, the so-called" reaction diffusion equations" have been used for many years for modeling a self-organized pattern far from equilibrium. In 1952, [73] showed that a motionless spatially periodic solution can be stable in a coupled reaction diffusion equations with two components if certain conditions are satisfied in the diffusion coefficients and the nonlinearity. It took about 40 years to observe the Turing pattern in real experiments. In a seminal paper, Castets and co-workers realized experimentally, for the first time, a Turing-type chemical pattern in a variant of chloride-iodide reaction in a gel reactor [7].

By changing the concentration of the chemical components, they realized that the spontaneous breaking of the translational symmetry is associated with the formation of a Turing structure [16]. Soon after, [60]-[61] carried out experiments with the same reaction and reactor more extended in the third direction to produce a stripe pattern and a hexagonal pattern depending on the concentrations of the chemical species. Theoretically, [73] have been studied numerically as an example of dissipative structures.

However, almost all of the previous investigations were restricted to one or two dimensions where only strip patterns, hexagonal patterns, and labyrinthine patterns could be realized. Only in some special cases some unusual patterns can appear. For example, the coupling between the different spatial modes in bi-stable systems causes resonant rhombic or quasi-periodic structures as has been predicted by the amplitude equations constructed by symmetric argument [18]. One may think of several reasons as to why three-dimensional numerical simulation of reaction diffusion systems is technically demanded and time consuming. It is also likely that three-dimensional Turing patterns were not so attractive because many researchers thought that two-dimensional patterns might be sufficient to understand the general properties of dissipative structures.

[1] used two numerical methods for investigating propagation heat solutions of PDEs, and used these methods to an autocatalytic reaction diffusion equations involving two diffusing chemicals in one-dimension.

[72] investigated responses of dynamic system to pulse perturbations theoretically and experimentally. Complex phenomena such as limit cycles, periodic solutions and chaos were numerically demonstrated.

[35] considered a discrete version of the Brusselator model of the famous Belousov-zhabotinsnky reaction in chemistry, and studied the dynamics of the local map. They discussed the set of trajectories that escape to infinity as well as analyzed the set of bounded trajectories.

[41]-[42] discussed that the formation of patterns and structures obtained through numerical simulation of the Turing mechanism in two-and three- dimensions. The forming patterns are found to depend strongly on the initial and boundary conditions as well as system parameters, showing a rich variety of patterns. [74] studied a second order method developed for the numerical solution of non-linear reaction diffusion equations in two space dimensions known as the "Brusselator" system. The algorithm is implemented in parallel using two processers, each solving a linear algebraic system as opposed to solving non-linear systems, which is often required for integrating non-linear partial differential equations(PDEs).

The work [29] is concerned with two aspects of pattern formation: pattern localization and degree of pattern order. In reaction-diffusion models, there are three major effects. These effects stem from the reaction terms, the diffusion terms and the presence or absence of precursor gradients. Both of these models have Hill coefficient 2; effects of higher cooperativities have recently been discussed.

[78] used lattice Boltzmann method to simulate a thermal flow which advents reactant species. As an example they considered a reactive system modeled by the Brusselator which exhibits Turing structures; the system is subjected to mixing as

induced by the wakes behind obstacles. The resulting effect is modification of the reaction patterns which can be as dramatic as their disappearance.

[64] considered a model for a differential – flow reactor based on cubic autocatalator kinetics in which the substrate is immobilized and the autocatalyst flows through the reactor at a constant rate. Linear stability analysis shows that there is a critical flow rate above which the spatially uniform steady state becomes convectively unstable. Numerical simulations show that this convective instability leads to the formation of a wave packet propagating through the reactor. The nature of this wave packet depends on the flow rate and on the two cases that are identified for the kinetic parameter μ, namely a generic case for general values of μ is close to the value which gives a Hopf bifurcation in the kinetic system.

[77] presented a fast and accurate numerical scheme for the solution of fifth-order boundary value problems with two point boundary conditions. The Adomian decomposition method and a modified form of this method are applied to construct the numerical solution. The new approach provides the solution in the form of rapidly convergent series and not at grid points. Two numerical illustrations are given to show the pertinent features of technique.

[49] studied some properties of the so called "canard solutions" that remain bounded in a full neighborhood of 0 and in the largest possible domain. The main goal is the complete asymptotic expansion of the difference between two values of the additional parameter corresponding to such solutions. For this purpose, they studied the behavior of the solutions near a turning point, they proved that for a large class of equations, if 0 is a turning point of order p, any solution y not exponentially large has, in some sector centered at 0, an asymptotic behavior (when e *approaches* 0) of the form $\sum Y_n(\frac{x}{e'})e'^n$, where $e'^{p+1} = e$ for $x = e'X$ with X large enough but independent of e. in the Brusselator case, moreover, a Stocks constant was computed for a particular nonlinear differential equation.

[38] considered the interaction of a small moving particle with a space-periodic pattern in a chemical reaction-diffusion system with a flow. The pattern is

produced by a one dimensional Brusselator model that is perturbed by a constant displacement from the equilibrium state at the inlet. By partially blocking the flow, the particle gives rise to a local increment of the flow rate. For certain parameter values, a response with intermittent Hopf and turning type structures is observed. In other regimes, a wave of substitution of missing peaks runs across the pattern.

[20] employed ADM to solve the momentum and energy equations for laminar boundary layer flow over flat plate at zero incidences with neglecting the frictional heating. A trial and error strategy was used to obtain the constant coefficient in the approximated solution. ADM provides an analytical solution in the form of an infinite power series. The effect of a domain polynomial terms is considered and shows that the accuracy of results is calculated. Also the effect of the Prandt1 number on the thermal boundary layer is obtained. The results show that ADM can solve the nonlinear differential equations with negligible error compared to the exact solution.

[65] showed that for exhibiting limit cycle behavior a two-component chemical reaction system has to involve at least three reactions among which one must be autocatalytic of the type 2X+…← → 3X+…. Under this condition, possible candidates for chemical limit cycle systems are selected by postulating their steady state to be an unstable focus. This procedure can be reduced to the selection of appropriate stoichiometric coefficients and is readily performed by a medium size computer. The result is quite a lot of limit cycle systems which are altogether simpler than, for example, "Brusselator" with its number of four reactions. One of the results is briefly discussed.

[56] investigated relaxation regimes for a doubly reduced Brusselator. Both P and *A* are generally, time dependent and have jumps in their time derivatives. The canonical forms of P are compared with the non-canonical ones in the context of robustness of the new concept with respect to incomplete information about the system studies. These forms of P are different in relaxation to a non-equilibrium state and coincide in relaxation to an equilibrium state.

[34] considered a discrete version of the Brusselator model of the famous Belousov-Zhabotinsky reaction in chemistry. The original model is a reaction-

diffusion equation and its discrete version is a coupled map lattice. Also they studied the dynamics of the local map, which is a smooth map of the plane, discussed the set of trajectories that escape to infinity and analyzed the set of bounded trajectories-the Julia set of the system.

[79] Oscillating Turing patterns arise in the system when bulk oscillations lose their stability to spatial perturbations. Spatially uniform external periodic forcing can generate oscillating Turing patterns when both the Turing and Hopf models are subcritical in the autonomous system. Most of the symmetric patterns show period doubling in both space and time. Patterns observed include squares, rhombi, stripes, and hexagons.

1.3 The Aim of the Thesis

1. To solve the Brusselator system in one dimensional space by:
   i) FDM (explicit, implicit) and FEM and we study their numerical stability.
   ii) WRM (Collocation, Galerkin, Least Square,Sub-Domain) methods.
   iii) Iterative Methods (SAM and MSAM)
2) In two dimensional space the system is solved by ADE and ADI methods also their stability is studied .
3) To make a comparison between these methods to find the most accurate one.

1.4 Layout of the Thesis:

This thesis contains six chapters:

Chapter One: In this chapter, some theoretical background of nonlinear partial differential equations, a general class of nonlinear-diffusion system, and some methods of their solution are given.

Chapter Two: In this chapter we use finite difference method, explicit, implicit (Crank-Nicolson) schemes to find the numerical solution of the Brusselator system in one dimensional space. The methods of ADE and ADI are also used to solve its two dimensional space.

Also, we study the numerical stability analysis of the system in one and two dimensional space; in one dimension, we study the stability of explicit, implicit schemes, and for two dimensions we study the stability of ADE and ADI schemes.

Chapter Three: In this chapter, Brusselator model is solved numerically by finite element method. The results obtained by finite element method are compared with those obtained by finite difference method.

Chapter Four: In this chapter, we use some expansion methods (Collocation method, Galerkin method, Least-square method and partition method) to solve Brusselator model numerically.

Chapter Five: In this chapter, iterative methods (successive approximation, and modified successive approximation) are used to approximate the solution of Brusselator system.

Chapter Six: In this chapter, three examples are presented and solved numerically to illustrate the efficiency of the presented method in this thesis and some conclusions and recommendations for further works are given.

CHAPTER TWO

Numerical Solution of Brusselator Model by Finite Difference Method and stability

2.1 Introduction:

Finite difference methods are found to be discrete techniques. When the domain of point interest is represented by a set of points or nodes and information between, these points are commonly obtained using Taylor series expansions [39].

The finite difference Scheme, generally, reduces a linear, nonlinear partial differential equations into system of linear , nonlinear equations and various methods were developed to find the numerical solution and acceleration the convergence [48].

Stability concepts are basic in many engineering and other applications. They are suggested by physician, where stability means, roughly speaking, that a small change (a small disturbance) of a physical system at some instant changes the behavior of the system only slightly at all future times t. Mathematically, stable means small perturbation in the initial data (or small error at any time) that remains small at later times. However, if small changes in the initial data produce large changes in the final results, the case will be unstable [4], [47].

In this chapter, we study the stability analysis of explicit and implicit methods for one and two dimensions with stability of Busselator medel.

2.2 Finite Difference Approximations

Let us consider the Taylor series of a function $y(x+\Delta x)$ evaluated around the point x, as

$$y(x+\Delta x) = y(x) + \Delta x\frac{dy(x)}{dx} + \frac{(\Delta x)^2}{2!}\frac{d^2y(x)}{dx^2} + \frac{(\Delta x)^3}{3!}\frac{d^3y(x)}{dx^3} + \dots.. \qquad (2.1)$$

Then for sufficiently small Δx, we can ignore the second and higher order terms, and so we have:

$$\frac{dy(x)}{dx} \approx \frac{y(x+\Delta x) - y(x)}{\Delta x}, \qquad (2.2)$$

which simply states that the slope of the function y at point x is approximated by the slope of the line segment joining between two points $(x, y(x))$ and $(x+\Delta x, y(x+\Delta x))$. Thus, the first derivative is calculated using the local discrete points, and this formula is correct to the first order since the first term omitted in Equation (2.1) is of order $(\Delta x)^2$. Similarly, if we expand the function $y(x-\Delta x)$ around the point x using the Taylor series method, we obtain the formula:

$$y(x-\Delta x) = y(x) - \Delta x \frac{dy(x)}{dx} + \frac{(\Delta x)^2}{2!}\frac{d^2 y(x)}{dx^2} - \frac{(\Delta x)^3}{3!}\frac{d^3 y(x)}{dx^3} + ... \qquad (2.3)$$

Again, when the second and higher order terms are ignored, the first derivative can be calculated from another formula using discrete points $(x, y(x))$ and $(x-\Delta x, y(x-\Delta x))$ as:

$$\frac{dy(x)}{dx} \approx \frac{y(x) - y(x-\Delta x)}{\Delta x}. \qquad (2.4)$$

Like Equation (2.2), this approximation of the first derivative at the point x requires only values close to that point. Another formula for approximating the first derivative can be obtained by subtracting Equation (2.1) from (2.3):

$$\frac{dy(x)}{dx} \approx \frac{y(x+\Delta x) - y(x-\Delta x)}{2\Delta x} \qquad (2.5)$$

This formula is more accurate than the last two (Equations (2.2) and (2.4)) because the first term truncated in deriving this equation containing $(\Delta x)^2$, compared to (Δx) in the last two formulas.

To obtain the discrete formula for the second derivative, we add Equation (2.1) and (2.3) and ignore the third order and higher order terms to obtain

$$\frac{d^2 y(x)}{dx^2} \approx \frac{y(x+\Delta x) - 2y(x) + y(x-\Delta x)}{(\Delta x)^2}. \qquad (2.6)$$

The first term truncated in deriving this formula contains $(\Delta x)^2$; hence, the error for this formula is comparable to Equation (2.5) obtained for the first derivative. We can proceed in this way to obtain higher order derivatives, but only the first and second derivatives are needed to solve the Brusselator model.

Let the domain [a, b] be divided into N equal intervals, the length of each interval is called the ***grid size*** given by :

$$\Delta x = \frac{b-a}{N}.$$

We define the point p as the point having the coordinate $p \cdot (\Delta x)$, and denote that point x_p; that is $x_0 = a, x_p = p \cdot (\Delta x) \quad for \quad p = 1, 2, ..., N, x_n = b.$

The variable y corresponding to the point x_p is denoted as y_p that is, $y(x_p) = y_p$. Equations (2.4,2.5 and 2.6) can be written as

$$(\frac{dy}{dx})_p \approx \frac{y_{p+1} - y_p}{\Delta x} \qquad (2.7a)$$

$$(\frac{dy}{dx})_p \approx \frac{y_p - y_{p-1}}{\Delta x} \qquad (2.7b)$$

$$(\frac{dy}{dx})_p \approx \frac{y_{p+1} - y_{p-1}}{2\Delta x} \qquad (2.7c)$$

respectively. The equations (2.7a, 2.7b and 2.7c) are called ***forward difference approximation***, ***backward difference approximation*** and ***central difference approximation*** *respectively.* The approximating formula for the second derivative is [62] :

$$(\frac{d^2 y}{dx^2})_p \approx \frac{y_{p+1} - 2y_p + y_{p-1}}{(\Delta x)^2} \qquad (2.8)$$

2.2.1 Grid Points

Let u be a function of independent variables x and t. The (x,t)-plane can be divided into a network of rectangles of sides $\Delta x = h$ and $\Delta t = k$ by drawing the set of lines :

$$\left. \begin{array}{ll} x_p = ph, & p = 0, 1, 2, 3, ... \\ t_q = qk, & q = 0, 1, 2, 3, ... \end{array} \right\} \qquad (2.9)$$

The point of intersection of these families of lines are called '**Grid Points**' (or sometimes referred to as **Lattice Points, Mesh Points**).

The general procedure to solve the partial differential equations by finite-difference approximation is to obtain the solution at these grid points.

In view of the lines $x = ph, t = qk$ defined above, we can rewrite the finite difference approximations to the first and second derivatives as follows:

The first order derivative of u with respect to x is given by:

$$u_x(x_p, t_q) = \frac{u_{p+1,q} - u_{p,q}}{h} + O(h) \quad (2.10)$$

$$u_x(x_p, t_q) = \frac{u_{p,q} - u_{p-1,q}}{h} + O(h) \quad (2.11)$$

$$u_x(x_p, t_q) = \frac{u_{p+1,q} - u_{p-1,q}}{2h} + O(h^2). \quad (2.12)$$

and

$$u_{xx}(x_p, t_q) = \frac{u_{p+1,q} - 2u_{p,q} + u_{p-1,q}}{h^2} + O(h^2) \quad (2.13)$$

Similarly, with respect to the independent variable t, we have

$$u_t(x_p, t_q) = \frac{u_{p,q+1} - u_{p,q}}{k} + O(k) \quad (2.14)$$

$$u_t(x_p, t_q) = \frac{u_{p,q} - u_{p,q-1}}{k} + O(k) \quad (2.15)$$

$$u_t(x_p, t_q) = \frac{u_{p,q+1} - u_{p,q-1}}{2k} + O(k^2) \quad (2.16)$$

And $u_{tt}(x_p, t_q) = \frac{u_{p,q+1} - 2u_{p,q} + u_{p,q-1}}{k^2} + O(k^2)$ (2.17)

Similar equations hold for v.

In the (x, t)-plane the above derivatives can be analyzed as follows [67] :

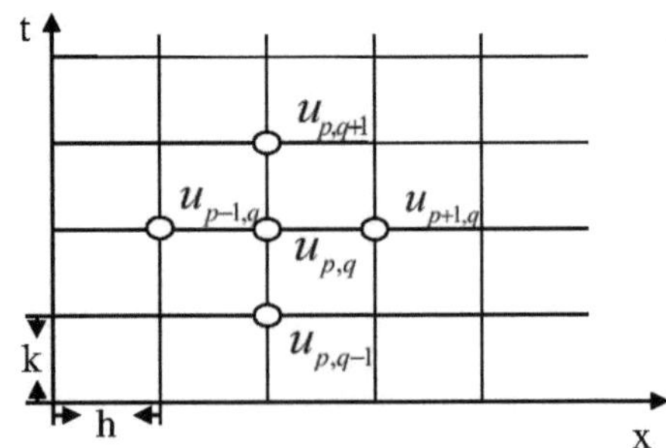

Fig. 2.1 : The grid point (mesh of explicit method) representation in xt-plane

2.2.2 Derivation of Explicit Method For Brusselator Model

Assume that the rectangle $R=\{(x,t):0\le x\le a,\quad 0\le t\le b\}$ is subdivided into *n*-1 by *m*-1 rectangle with sizes $\Delta x=h$ and $\Delta t=k$, as shown in Figure 2.1 . Start at the bottom row, where $t=t_1=0$, and the solution is $u(x_p,t_1)=f(x_p)$. A method is used for computing the approximations to $u(x,t)$ at grid points in successive rows $\{u(x_p,t_q):p=1,2,...,n\}$, for $q=2,3,...,m$. The different formulas used for approximation $u_t(x,t), u_x(x,t)$ *and* $u_{xx}(x,t)$ are

$$u_t(x,t)=\frac{u(x,t+k)-u(x,t)}{k}+O(k) \tag{2.18}$$

$$u_x(x,t)=\frac{u(x+h,t)-u(x,t)}{h}+O(h) \tag{2.19}$$

and

$$u_{xx}(x,t)=\frac{u(x-h,t)-2u(x,t)+u(x+h,t)}{h^2}+O(h^2) \tag{2.20}$$

The grid spacing is uniform in every row: $x_{p+1}=x_p+h$ and ($x_{p-1}=x_p-h$), and it is uniform in every column: $t_{q+1}=t_q+k$ and ($t_{q-1}=t_q-k$). Next, we drop the terms $O(k), O(h)$ and $O(h^2)$ [48] and use the approximation $u_{p,q}$ for $u(x_p,t_q)$ in equation (2.18), (2.19) and substitute into Brusselator model (1.1) to obtain:

$$\begin{aligned}\frac{u_{p,q+1}-u_{p,q}}{k}&=d_1\frac{u_{p+1,q}-2u_{p,q}+u_{p-1,q}}{h^2}-(b+1)u_{p,q}+u_{p,q}^2v_{p,q}+a,\\ \frac{v_{p,q+1}-v_{p,q}}{k}&=d_2\frac{v_{p+1,q}-2v_{p,q}+v_{p-1,q}}{h^2}-u_{p,q}^2v_{p,q}+bu_{p,q}.\end{aligned} \tag{2.21}$$

$$u_{p,q+1}=\frac{d_1k}{h^2}(u_{p+1,q}-2u_{p,q}+u_{p-1,q})+ku_{p,q}^2v_{p,q}-k(b+1)u_{p,q}+u_{p,q}+ka, \tag{2.22}$$

$$v_{p,q+1}=\frac{d_2k}{h^2}(v_{p+1,q}-2v_{p,q}+v_{p-1,q})-ku_{p,q}^2v_{p,q}+kbu_{p,q}.$$

and

$$u_{p,q+1}=\frac{d_1k}{h^2}(u_{p+1,q}+u_{p-1,q})+(-2\frac{d_1k}{h^2}-k(b+1)+1)u_{p,q}+ku_{p,q}^2v_{p,q}+ka,$$
$$v_{p,q+1}=\frac{d_2k}{h^2}(v_{p+1,q}+v_{p-1,q})+(-2\frac{d_2k}{h^2}-ku_{p,q}^2+1)v_{p,q}+kbu_{p,q}. \quad (2.23)$$

Let $\frac{d_1k}{h^2}=r_1$ and $\frac{d_2k}{h^2}=r_2$, then

$$u_{p,q+1}=r_1(u_{p+1,q}+u_{p-1,q})+(1-2r_1-k(b+1))u_{p,q}+ku_{p,q}^2v_{p,q}+ka,$$
$$v_{p,q+1}=r_2(v_{p+1,q}+v_{p-1,q})+(1-2r_2-ku_{p,q}^2)v_{p,q}+kbu_{p,q}. \quad (2.24)$$

The result is the explicit forward difference equation to the Brusselator model.

2.2.3 Derivation of Implicit (Crank-Nicholson) Method For Brusselator Model

This method was invented by Crank and Nicholson in (1947) based on numerical approximations for solutions. They replaced u_{xx} by the mean of its finite difference representation of the $(q)^{th}$ and $(q+1)^{th}$ time rows as shown in the Fig. 2.2 [69].

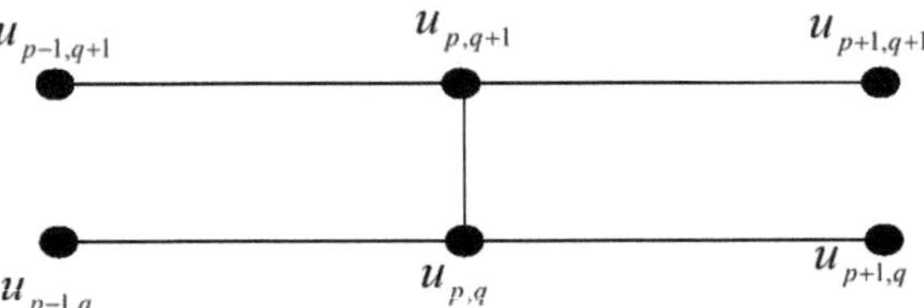

Fig. 2.2 The mesh of Crank- Nicholson Method

Approximating the system of the Brusselator model, where

$$u_{xx}=\frac{1}{2}(\frac{u_{p+1,q}-2u_{p,q}+u_{p-1,q}}{h^2}+\frac{u_{p+1,q+1}-2u_{p,q+1}+u_{p-1,q+1}}{h^2}),$$

and (2.25)

$$v_{xx}=\frac{1}{2}(\frac{v_{p+1,q}-2v_{p,q}+v_{p-1,q}}{h^2}+\frac{v_{p+1,q+1}-2v_{p,q+1}+v_{p-1,q+1}}{h^2}).$$

Substituting it in the Brusselator model (1.1) gives:

$$\frac{u_{p,q+1}-u_{p,q}}{k}=\frac{d_1}{2h^2}(u_{p+1,q}-2u_{p,q}+u_{p-1,q}+u_{p+1,q+1}-2u_{p,q+1}+u_{p-1,q+1})-(b+1)u_{p,q}+u_{p,q}^2v_{p,q}+a,$$

(2.26)

$$\frac{v_{p,q+1}-v_{p,q}}{k}=\frac{d_2}{2h^2}(v_{p+1,q}-2v_{p,q}+v_{p-1,q}+v_{p+1,q+1}-2v_{p,q+1}+v_{p-1,q+1})+bu_{p,q}-u_{p,q}^2v_{p,q}.$$ Let $r_1=\dfrac{d_1k}{2h^2}$, and

$r_2=\dfrac{d_2k}{2h^2}$, so

$$u_{p,q+1}=r_1(u_{p+1,q}-2u_{p,q}+u_{p-1,q}+u_{p+1,q+1}-2u_{p,q+1}+u_{p-1,q+1})-k(b+1)u_{p,q}+ku_{p,q}^2v_{p,q}+u_{p,q}+ka,$$ and

(2.27)

$$v_{p,q+1}=r_2(v_{p+1,q}-2v_{p,q}+v_{p-1,q}+v_{p+1,q+1}-2v_{p,q+1}+v_{p-1,q+1})+v_{p,q}+kbu_{p,q}+ku_{p,q}^2v_{p,q}.$$

Rearranging the last two equations gives :

$$(1+2r_1)u_{p,q+1}-r_1(u_{p+1,q+1}+u_{p-1,q+1})=(1-k(b+1)-2r_1)u_{p,q}+r_1(u_{p-1,q}+u_{p+1,q})+ku_{p,q}^2v_{p,q}+ka,$$

and (2.28)

$$(1+2r_2)v_{p,q+1}-r_2(v_{p+1,q+1}+v_{p-1,q+1})=(1-2r_2)v_{p,q}+kbu_{p,q}+ku_{p,q}^2v_{p,q}+r_2(v_{p+1,q}+v_{p-1,q}).$$

This system of equations (2.28) represents the implicit difference approximation for Brusselator model, where the left hand side of the system contains three unknown values, while the right hand side contains three known values for p=2,3,…,n-1 .

The above mentioned system represent the crank-Nicolson representation to the Brusselator model. This time we must solve the system (2.28) for the three values $u_{p-1,q+1}, u_{p,q+1}$ and $u_{p+1,q+1}$ for p=2,3,...,n-1.The terms on the right-hand side of equation (2.28) are all known. Hence the first equation in (2.28) forms a tri-diagonal linear system in the form:

$$\boldsymbol{AU=B} \qquad (2.29)$$

The boundary conditions are used in the first and last equations (i.e. $u_{1,q}=u_{n,q}=0$ and $u_{1,q+1}=u_{n,q+1}=0$ respectively). The equation (2.28) is especially pleasing to view in their tri-diagonal matrix form **(2.29)** .

$$
\begin{bmatrix}
(1+2r_1) & -r_1 & 0 & 0 & 0 & . & . & . & 0 \\
-r_1 & (1+2r_1) & -r_1 & 0 & 0 & 0 & 0 & 0 & 0 \\
0 & -r_1 & (1+2r_1) & -r_1 & 0 & 0 & 0 & 0 & 0 \\
0 & 0 & -r_1 & (1+2r_1) & -r_1 & & & & \\
. & 0 & 0 & -r_1 & . & . & & & \\
. & . & 0 & 0 & . & . & . & & \\
. & . & . & 0 & 0 & . & . & . & \\
. & . & . & . & . & . & . & . & . \\
. & . & . & . & . & . & . & . & . \\
0 & 0 & 0 & 0 & 0 & 0 & -r_1 & (1+2r_1) & -r_1 \\
0 & 0 & 0 & 0 & 0 & 0 & 0 & -r_1 & (1+2r_1)
\end{bmatrix}
\begin{bmatrix}
u_{2,q+1} \\ u_{3,q+1} \\ u_{4,q+1} \\ u_{5,q+1} \\ . \\ . \\ . \\ \\ u_{n-3,q+1} \\ u_{n-2,q+1} \\ u_{n-1,q+1}
\end{bmatrix}
=
$$

$$
\square
\begin{bmatrix}
r_1 u_{3,q} + (1-2r_1 - k(b+1))u_{2,q} + k{u^2}_{2,q} v_{2,q} + ka \\
r_1(u_{4,q} + u_{2,q}) + (1-2r_1 - k(b+1))u_{3,2} + k u_{3,q}^2 v_{3,q} + ka \\
r_1(u_{5,q} + u_{3,q}) + (1-2r_1 - k(b+1))u_{4,q} + k u_{4,q}^2 v_{4,q} + ka \\
. \\
. \\
. \\
. \\
r_1(u_{n-2,q} + u_{n-4,q}) + (1-2r_1 - k(b+1))u_{n-3,q} + k u_{n-3,q}^2 v_{4,q} + ka \\
r_1(u_{n-1,q} + u_{n-3,q}) + (1-2r_1 - k(b+1))u_{n-2,q} + k u_{n-2,q}^2 v_{n-2,q} + ka \\
\\
r_1 u_{n-2,q} + (1-2r_1 - k(b+1))u_{n-1,q} + k{u^2}_{n-1,q} v_{n-1,q} + ka
\end{bmatrix}
\qquad (2.30)
$$

Similarly from the second equation of (2.28) we obtian a tridiagonal linear system . For p=2,3,4,…,n-1, $v_{1,q} = v_{n,q} = 0$ and $v_{1,q+1} = v_{n,q+1} = 0$

$$
\begin{bmatrix}
(1+2r_1) & -r_1 & 0 & 0 & 0 & . & . & . & 0 \\
-r_1 & (1+2r_1) & -r_1 & 0 & 0 & 0 & 0 & 0 & 0 \\
0 & -r_1 & (1+2r_1) & -r_1 & 0 & 0 & 0 & 0 & 0 \\
0 & 0 & -r_1 & (1+2r_1) & -r_1 & & & & \\
. & 0 & 0 & -r_1 & . & . & & & \\
. & . & 0 & 0 & . & . & . & & \\
. & . & . & 0 & 0 & . & . & . & \\
. & . & . & . & . & . & . & . & . \\
. & . & . & . & . & . & . & . & . \\
0 & 0 & 0 & 0 & 0 & 0 & -r_1 & (1+2r_1) & -r_1 \\
0 & 0 & 0 & 0 & 0 & 0 & 0 & -r_1 & (1+2r_1)
\end{bmatrix}
\begin{bmatrix}
v_{2,q+1} \\ v_{3,q+1} \\ v_{4,q+1} \\ v_{5,q+1} \\ . \\ . \\ . \\ \\ v_{n-3,q+1} \\ v_{n-2,q+1} \\ v_{n-1,q+1}
\end{bmatrix} =
$$

$$
\begin{bmatrix}
r_2 v_{3,q} + (1-2r_2)u_{2,q} - ku^2{}_{2,q} v_{2,q} + kbu_{2,q} \\
r_2(v_{4,q} + v_{2,q}) + (1-2r_2)v_{3,2} - ku^2_{3,q} v_{3,q} + kbu_{3,q} \\
r_2(v_{5,q} + v_{3,q}) + (1-2r_2)v_{4,q} - ku^2_{4,q} v_{4,q} + kbu_{4,q} \\
. \\
. \\
. \\
. \\
r_2(v_{n-2,q} + v_{n-4,q}) + (1-2r_2)v_{n-3,q} - ku^2_{n-3,q} v_{n-3,q} + kbu_{n-3,q} \\
r_2(v_{n-1,q} + v_{n-3,q}) + (1-2r_2)v_{n-2,q} - ku^2_{n-2,q} v_{n-2,q} + kbu_{n-2,q} \\
\\
r_2 v_{n-2,q} + (1-2r_2)v_{n-1,q} - ku^2{}_{n-1,q} v_{n-1,q} + kbu_{n-1,q}
\end{bmatrix}. \tag{2.31}
$$

When the Crank-Nicolson method is implemented with a computer program, the linear system ***AV=B*** can be solved by either direct methods or by iterations method. In this study, we use the direct Gaussian elimination method [8] to solve the linear system in (2.31).

2.3Two Dimensional Space

2.3.1: Derivation of Alternating Direction Explicit (ADE)

The two dimensional Brusselator model is given by

$$\frac{\partial u}{\partial t}=d_1(\frac{\partial^2 u}{\partial x^2}+\frac{\partial^2 u}{\partial y^2})+u^2 v-(b+1)u+a, \quad (t,x,y)\in(0,\infty)\times\Omega$$

$$\frac{\partial v}{\partial t}=d_2(\frac{\partial^2 v}{\partial x^2}+\frac{\partial^2 v}{\partial y^2})-u^2 v+bu, \qquad (t,x,y)\in(0,\infty)\times\Omega$$

Neumann B.C.

$$\frac{\partial u}{\partial x}=\frac{\partial u}{\partial y}=0, \ at\ x,y=0, and\ x,y=L$$

$$\frac{\partial v}{\partial x}=\frac{\partial v}{\partial y}=0, \ at\ x,y=0, and\ x,y=L$$

I.C. and B.C.

$$u(0,x,y)=u_0(x,y),\ v(0,x,y)=v_0(x,y), x,y\in\partial\Omega$$

$$u(t,x,y)=v(t,x,y)=0,\ t>0, x,y\in\partial\Omega$$

This method is referred to as alternating direction explicit since a single cycle of computation requires the solution of the two different finite difference approximations written in different physical directions. The end result of the two cycles is taken as the answer of the (n+1) plane [39].

We consider a square region $0\leq x\leq 1$, $0\leq y\leq 1$ and u, v is known at all points within and on the boundary of the square region. We draw lines parallel to x,y,t – axis as $x_p=ph$, $y_q=qk$, and $t_n=nz$, $p,q=0,1,2,\ldots,M$ and $n=0,1,2,\ldots,N$, where $h=\delta x$, $k=\delta y$, $z=\delta t$.

Denote the values of u at these mesh points by $u(ph,qk,nz)=u_{p,q,n}$ and $v(ph,qk,nz)=v_{p,q,n}$. The explicit finite difference representation of the Brusselator model is (assuming that $h=k$):

$$\frac{u_{p,q,n+1}-u_{p,q,n}}{z}=\frac{d_1}{h^2}(u_{p+1,q,n}-2u_{p,q,n}+u_{p-1,q,n}+u_{p,q+1,n}-2u_{p,q,n}+u_{p,q-1,n})-(b+1)u_{p,q,n}+u^2_{p,q,n}v_{p,q,n}+a$$

$$\frac{v_{p,q,n+1}-v_{p,q,n}}{z}=\frac{d_2}{h^2}(v_{p+1,q,n}-2v_{p,q,n}+v_{p-1,q,n}+v_{p,q+1,n}-2v_{p,q,n}+v_{p,q-1,n})+bu_{p,q,n}-u^2_{p,q,n}v_{p,q,n}$$

and

$$u_{p,q,n+1}-u_{p,q,n}=\frac{d_1 z}{h^2}(u_{p+1,q,n}-2u_{p,q,n}+u_{p-1,q,n}+u_{p,q+1,n}-2u_{p,q,n}+u_{p,q-1,n})-z(b+1)u_{p,q,n}+zu^2_{p,q,n}v_{p,q,n}+az$$

$$v_{p,q,n+1}-v_{p,q,n}=\frac{d_2 z}{h^2}(v_{p+1,q,n}-2v_{p,q,n}+v_{p-1,q,n}+v_{p,q+1,n}-2v_{p,q,n}+v_{p,q-1,n})+bzu_{p,q,n}-zu^2_{p,q,n}v_{p,q,n}$$

Let $m_i = \frac{d_i z}{h^2}, i = 1,2.$ Then we simplify the system to obtain:

$$u_{p,q,n+1} - u_{p,q,n} = m_1(u_{p+1,q,n} - 2u_{p,q,n} + u_{p-1,q,n} + u_{p,q+1,n} - 2u_{p,q,n} + u_{p,q-1,n}) - z(b+1)u_{p,q,n} + zu^2_{p,q,n}v_{p,q,n} + az$$

$$v_{p,q,n+1} - v_{p,q,n} = m_2(v_{p+1,q,n} - 2v_{p,q,n} + v_{p-1,q,n} + v_{p,q+1,n} - 2v_{p,q,n} + v_{p,q-1,n}) + bzu_{p,q,n} - zu^2_{p,q,n}v_{p,q,n}$$

and

$$u_{p,q,n+1} = (1 - 4m_1 - z(b+1))u_{p,q,n} + m_1(u_{p+1,q,n} + u_{p-1,q,n} + u_{p,q+1,n} + u_{p,q-1,n}) + zu^2_{p,q,n}v_{p,q,n} + az$$

$$v_{p,q,n+1} = (1 - 4m_2)v_{p,q,n} + m_2(v_{p+1,q,n} + v_{p-1,q,n} + v_{p,q+1,n} + v_{p,q-1,n}) + bzu_{p,q,n} - zu^2_{p,q,n}v_{p,q,n}$$

This is the alternating direction explicit formula for the Brusselator model:

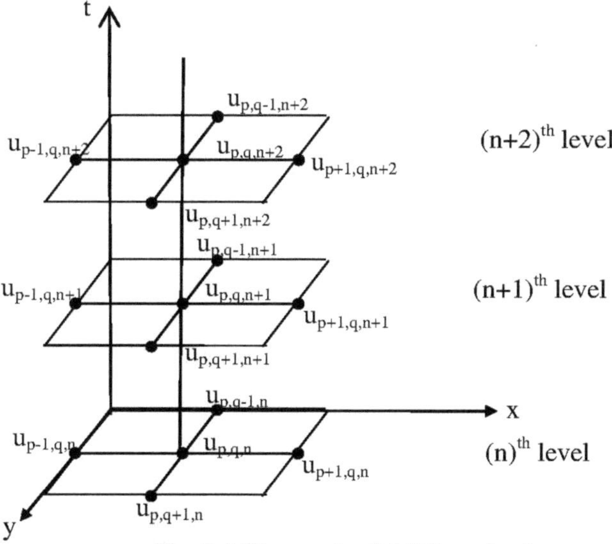

Fig. 2.3 The mesh of ADE method

2.3.2 Derivation of Alternating Direction Implicit (ADI)

This method is developed by Peaceman and Rachford [56], and is called Alternating Direction Implicit (ADI) method. This two-step approach requires minimal computer storage and is quite accurate than ADE. Further, ADI method is unconditionally stable. The method involves the alternate of two different finite difference approximations to the two dimensional space [2].

In the ADI approach, the finite difference equations are written in terms of quantities at two x levels. However, two different finite difference approximations are used alternately, one to advance the calculations from the plane (n) to a plane $(n+1)$, and the second to advance the calculations from $(n+1)$ plane to the $(n+2)$ plane [39]. With this method, each of the two steps involves diffusion in both the x and y-directions. In the first step the diffusion in x is modeled implicitly with diffusion in y model explicitly, with the roles reversed in the second step [43], then we advance the solution of Brusselator model from nth plane to $(n+1)^{th}$ plane by replacing $\frac{\partial^2 u}{\partial x^2}$ and $\frac{\partial^2 v}{\partial x^2}$ by implicit finite difference approximation at the (n)th plane, with these approximations , from Brusselator model, we get:

$$\frac{u_{p,q,n+1}-u_{p,q,n}}{z}=\frac{d_1}{h^2}(u_{p+1,q,n+1}-2u_{p,q,n+1}+u_{p-1,q,n+1})+\frac{d_1}{k^2}(u_{p,q+1,n}-2u_{p,q,n}+u_{p,q-1,n})-(b+1)u_{p,q,n}+u^2_{p,q,n}v_{p,q,n}+a,$$

$$\frac{v_{p,q,n+1}-v_{p,q,n}}{z}=\frac{d_2}{h^2}(v_{p+1,q,n+1}-2v_{p,q,n+1}+v_{p-1,q,n+1})+\frac{d_2}{K^2}(v_{p,q+1,n}-2v_{p,q,n}+v_{p,q-1,n})+bu_{p,q,n}-u^2_{p,q,n}v_{p,q,n},$$

and

$$\frac{u_{p,q,n+2}-u_{p,q,n+1}}{z}=\frac{d_1}{h^2}(u_{p+1,q,n+1}-2u_{p,q,n+1}+u_{p-1,q,n+1})+\frac{d_1}{k^2}(u_{p,q+1,n+2}-2u_{p,q,n+2}+u_{p,q-1,n+2})-(b+1)u_{p,q,n}+u^2_{p,q,n}v_{p,q,n}+a,$$

$$\frac{v_{p,q,n+2}-v_{p,q,n+1}}{z}=\frac{d_2}{h^2}(v_{p+1,q,n+1}-2v_{p,q,n+1}+v_{p-1,q,n+1})+\frac{d_2}{k^2}(v_{p,q+1,n+2}-2v_{p,q,n+2}+v_{p,q-1,n+2})+bu_{p,q,n}-u^2_{p,q,n}v_{p,q,n}.$$

Simplifying the system will give:

$$u_{p,q,n+1}=\frac{d_1 z}{h^2}(u_{p+1,q,n+1}-2u_{p,q,n+1}+u_{p-1,q,n+1})+\frac{d_1 z}{k^2}(u_{p,q+1,n}-2u_{p,q,n}+u_{p,q-1,n})-z(b+1)u_{p,q,n}+zu^2_{p,q,n}v_{p,q,n}+az+u_{p,q,n}$$

$$v_{p,q,n+1}=\frac{d_2 z}{h^2}(v_{p+1,q,n+1}-2v_{p,q,n+1}+v_{p-1,q,n+1})+\frac{d_2 z}{k^2}(v_{p,q+1,n}-2v_{p,q,n}+v_{p,q-1,n})+zbu_{p,q,n}-zu^2_{p,q,n}v_{p,q,n}+v_{p,q,n+1}.$$

From the second two equations we will obtain a system of the form

$$u_{p,q,n+2}=\frac{d_1 z}{h^2}(u_{p+1,q,n+1}-2u_{p,q,n+1}+u_{p-1,q,n+1})+\frac{d_1 z}{k^2}(u_{p,q+1,n+2}-2u_{p,q,n+2}+u_{p,q-1,n+2})-z(b+1)u_{p,q,n}+zu^2_{p,q,n}v_{p,q,n}+az+u_{p,q,n+1}$$

$$v_{p,q,n+2}=\frac{d_2 z}{h^2}(v_{p+1,q,n+1}-2v_{p,q,n+1}+v_{p-1,q,n+1})+\frac{d_2 z}{k^2}(v_{p,q+1,n+2}-2v_{p,q,n+2}+v_{p,q-1,n+2})+zbu_{p,q,n}-zu^2_{p,q,n}v_{p,q,n}+v_{p,q,n+1}.$$

From the first two equations where $r_1=\frac{d_1 z}{h^2}$ *and* $r_2=\frac{d_1 z}{k^2}$ and $m_1=\frac{d_2 z}{h^2}$, and $m_2=\frac{d_2 z}{k^2}$,the two systems will take the form:

$$u_{p,q,n+1}=r_1(u_{p+1,q,n+1}-2u_{p,q,n+1}+u_{p-1,q,n+1})+r_2(u_{p,q+1,n}-2u_{p,q,n}+u_{p,q-1,n})-z(b+1)u_{p,q,n}+zu^2_{p,q,n}v_{p,q,n}+az+u_{p,q,n}$$

$$v_{p,q,n+1}=m_1(v_{p+1,q,n+1}-2v_{p,q,n+1}+v_{p-1,q,n+1})+m_2(v_{p,q+1,n}-2v_{p,q,n}+v_{p,q-1,n})+zbu_{p,q,n}-zu^2_{p,q,n}v_{p,q,n}+v_{p,q,n}$$

and this implies that

$$u_{p,q,n+1} = r_1(u_{p+1,q,n+1} - 2u_{p,q,n+1} + u_{p-1,q,n+1}) + r_2(u_{p,q+1,n} + u_{p,q-1,n}) + (1 - 2r_1 - z(b+1)u_{p,q,n} + zu^2_{p,q,n}v_{p,q,n} + az$$

$$v_{p,q,n+1} = m_1(v_{p+1,q,n+1} - 2v_{p,q,n+1} + v_{p-1,q,n+1}) + m_2(v_{p,q+1,n} + v_{p,q-1,n}) + (1 - 2m_2)v_{p,q,n} + zbu_{p,q,n} - zu^2_{p,q,n}v_{p,q,n}$$

and also simplifying these systems of equations yields:

$$(1+2r_1)u_{p,q,n+1} = r_1(u_{p+1,q,n+1} + u_{p-1,q,n+1}) + r_2(u_{p,q+1,n} + u_{p,q-1,n}) + (1 - 2r_2 - z(b+1))u_{p,q,n} + az + zu^2_{p,q,n}v_{p,q,n}$$

$$(1+2m_1)v_{p,q,n+1} = m_1(v_{p+1,q,n+1} + v_{p-1,q,n+1}) + m_2(v_{p,q+1,n} + v_{p,q-1,n}) + (1 - 2m_2)v_{p,q,n} + bzu_{p,q,n} - zu^2_{p,q,n}v_{p,q,n}$$

also the second system of equations is :

$$(1+2r_2)u_{p,q,n+2} = r_1(u_{p+1,q,n+1} + u_{p-1,q,n+1}) + (1-2r_1)u_{p,q,n+1} + r_2(u_{p,q+1,n+2} + u_{p,q-1,n+2}) - z(b+1)u_{p,q,n} + zu^2_{p,q,n}v_{p,q,n} + az$$

$$(1+2m_2)v_{p,q,n+2} = m_1(v_{p+1,q,n+1} + v_{p-1,q,n+1}) + (1-2m_1)v_{p,q,n+1} + m_2(v_{p,q+1,n+2} + v_{p,q-1,n+2}) + zbu_{p,q,n} - zu^2_{p,q,n}v_{p,q,n}$$

The last two systems represent Alternating Direction implicit under the conditions :

$u_{1,q,n+1} = u_{1,q+1,n+1} = 0$

$u_{M,q,n+1} = u_{M,q+1,n+1} = 0.$

The tri-diagonal matrices for the system in the level n advanced to the level $(n+1)$, for u can be formulated as follows $AU=B$.

$$\begin{bmatrix}
(1+2r_1) & -r_1 & 0 & 0 & 0 & . & . & . & 0 \\
-r_1 & (1+2r_1) & -r_1 & 0 & 0 & 0 & 0 & 0 & 0 \\
0 & -r_1 & (1+2r_1) & -r_1 & 0 & 0 & 0 & 0 & 0 \\
0 & 0 & -r_1 & (1+2r_1) & -r_1 & & & & \\
. & 0 & 0 & -r_1 & . & . & & & \\
. & . & 0 & 0 & . & . & . & & \\
. & . & . & 0 & 0 & . & . & . & \\
. & . & . & . & . & . & . & . & . \\
. & . & . & . & . & . & . & . & . \\
0 & 0 & 0 & 0 & 0 & 0 & -r_1 & (1+2r_1) & -r_1 \\
0 & 0 & 0 & 0 & 0 & 0 & 0 & -r_1 & (1+2r_1)
\end{bmatrix}
\begin{bmatrix}
u_{2,q,n+1} \\ u_{3,q,n+1} \\ u_{4,q,n+1} \\ u_{5,q,n+1} \\ . \\ . \\ . \\ u_{n-3,q,n+1} \\ u_{n-2,q,n+1} \\ u_{M-1,q,n+1}
\end{bmatrix}$$

$$
=\begin{bmatrix}
az+r_2(u_{2,q+1,n}+u_{2,q-1,n})+(1-2r_2-z(b+1))u_{2,q,n}+zu^2{}_{2,q,n}v_{2,q,n} \\
az+r_2(u_{3,q+1,n}+u_{3,q-1,n})+(1-2r_2-z(b+1))u_{3,q,n}+zu^2{}_{3,q,n}v_{3,q,n} \\
az+r_2(u_{4,q+1,n}+u_{4,q-1,n})+(1-2r_2-z(b+1))u_{4,q,n}+zu^2{}_{4,q,n}v_{4,q,n} \\
. \\
. \\
. \\
. \\
az+r_2(u_{M-1,q+1,n}+u_{M-1,q-1,n})+(1-2r_2-z(b+1))u_{M-1,q,n}+zu^2{}_{M-1,q,n}v_{M-1,q,n}
\end{bmatrix}.
$$

And for the system $AV=B$ the tri-diagonal is in the form:

$$
\begin{bmatrix}
(1+2m_1) & -m_1 & 0 & 0 & 0 & . & . & . & 0 \\
-m_1 & (1+2m_1) & -m_1 & 0 & 0 & 0 & 0 & 0 & 0 \\
0 & -m_1 & (1+2m_1) & -m_1 & 0 & 0 & 0 & 0 & 0 \\
0 & 0 & -m_1 & (1+2m_1) & -m_1 & & & & \\
. & 0 & 0 & -m_1 & . & . & & & \\
. & . & 0 & 0 & . & . & . & & \\
. & . & . & 0 & 0 & . & . & . & \\
. & . & . & . & . & . & . & . & . \\
. & . & . & . & . & . & . & . & . \\
0 & 0 & 0 & 0 & 0 & 0 & -m_1 & (1+2m_1) & -m_1 \\
0 & 0 & 0 & 0 & 0 & 0 & 0 & -m_1 & (1+2m_1)
\end{bmatrix}
\begin{bmatrix}
v_{2,q,n+1} \\ v_{3,q,n+1} \\ v_{4,q,n+1} \\ v_{5,q,n+1} \\ . \\ . \\ . \\ v_{M-3,q,n+1} \\ v_{M-2,q,n+1} \\ v_{M-1,q,n+1}
\end{bmatrix}
$$

$$
=\begin{bmatrix}
m_2(v_{2,q+1,n}+v_{2,q-1,n})+(1-2m_2)v_{2,q,n}+zbu_{2,q,n}-zu^2{}_{2,q,n}v_{2,q,n} \\
m_2(v_{3,q+1,n}+v_{3,q-1,n})+(1-2m_2)v_{3,q,n}+zbu_{3,q,n}-zu^2{}_{3,q,n}v_{3,q,n} \\
m_2(v_{4,q+1,n}+v_{4,q-1,n})+(1-2m_2)v_{4,q,n}+zbu_{4,q,n}-zu^2{}_{4,q,n}v_{4,q,n} \\
. \\
. \\
. \\
. \\
m_2(v_{M-1,q+1,n}+v_{M-1,q-1,n})+(1-2m_2)v_{M-1,q,n}+zbu_{M-1,q,n}-zu^2{}_{M-1,q,n}v_{M-1,q,n}
\end{bmatrix}.
$$

And the tri-diagonal matrices for the system in level $(n+1)$ advanced to level $(n+2)$ for both u is given by $AU=B$:

$$
\begin{bmatrix}
(1+2r_2) & -r_2 & 0 & 0 & 0 & . & . & . & 0 \\
-r_2 & (1+2r_2) & -r_2 & 0 & 0 & 0 & 0 & 0 & 0 \\
0 & -r_1 & (1+2r_1) & -r_1 & 0 & 0 & 0 & 0 & 0 \\
0 & 0 & -r_2 & (1+2r_2) & -r_2 & & & & \\
. & 0 & 0 & -r_2 & . & . & & & \\
. & . & 0 & 0 & . & . & . & & \\
. & . & . & 0 & 0 & . & . & . & \\
. & . & . & . & . & . & . & . & . \\
. & . & . & . & . & . & . & . & . \\
0 & 0 & 0 & 0 & 0 & 0 & -r_2 & (1+2r_2) & -r_2 \\
0 & 0 & 0 & 0 & 0 & 0 & 0 & -r_2 & (1+2r_2)
\end{bmatrix}
\begin{bmatrix}
u_{p,2,n+2} \\ u_{p,3,n+2} \\ u_{p,4,n+2} \\ u_{p,5,n+2} \\ . \\ . \\ . \\ \\ u_{p,M-3,n+2} \\ u_{p,M-2,n+2} \\ u_{p,M-1,n+2}
\end{bmatrix}
=
$$

$$
\begin{bmatrix}
az + r_1(u_{p+1,2,n+1} + u_{p-1,2,n+1}) + (1-2r_1)u_{p,2,n+1} - z(b+1))u_{p,2,n} + zu^2{}_{p,2,n}v_{p,2,n} \\
az + r_1(u_{p+1,3,n+1} + u_{p-1,3,n+1}) + (1-2r_1)u_{p,3,n+1} - z(b+1))u_{p,3,n} + zu^2{}_{p,3,n}v_{p,3,n} \\
az + r_1(u_{p+1,4,n+1} + u_{p-1,4,n+1}) + (1-2r_1)u_{p,4,n+1} - z(b+1))u_{p,4,n} + zu^2{}_{p,4,n}v_{p,4,n} \\
az + r_1(u_{p+1,5,n+1} + u_{p-1,5,n+1}) + (1-2r_1)u_{p,5,n+1} - z(b+1))u_{p,5,n} + zu^2{}_{p,5,n}v_{p,5,n} \\
. \\ . \\ . \\ . \\
\\
az + r_1(u_{p+1,M-1,n+1} + u_{p-1,M-1,n+1}) + (1-2r_1)u_{p,M-1,n+1} - z(b+1))u_{pM-1,2,n} + zu^2{}_{p,2,n}v_{pM-1,n+1}
\end{bmatrix}
$$

Also for $AV=B$ the tri-diagonal is in the form :

$$
\begin{bmatrix}
(1+2m_2) & -m_2 & 0 & 0 & 0 & . & . & . & 0 \\
-m_2 & (1+2m_2) & -m_2 & 0 & 0 & 0 & 0 & 0 & 0 \\
0 & -m_2 & (1+2m_2) & -m_2 & 0 & 0 & 0 & 0 & 0 \\
0 & 0 & -m_2 & (1+2m_2) & -m_2 & & & & \\
. & 0 & 0 & -m_1 & . & . & & & \\
. & . & 0 & 0 & . & . & . & & \\
. & . & . & 0 & 0 & . & . & . & \\
. & . & . & . & . & . & . & . & . \\
. & . & . & . & . & . & . & . & . \\
0 & 0 & 0 & 0 & 0 & 0 & -m_2 & (1+2m_2) & -m_2 \\
0 & 0 & 0 & 0 & 0 & 0 & 0 & -m_2 & (1+2m_2)
\end{bmatrix}
\begin{bmatrix}
v_{p,2,n+2} \\ v_{p,3,n+2} \\ v_{p,4,n+2} \\ v_{p,5,n+2} \\ . \\ . \\ . \\ \\ v_{p,M-3,n+2} \\ v_{p,M-2,n+2} \\ v_{p,M-1,n+2}
\end{bmatrix}
=
\begin{bmatrix}
m_1(v_{p+1,2,n+1}+v_{p-1,2,n+1})+(1-2m_1)v_{p,2,n+1}+zbu_{p,2,n}-zu^2{}_{p,2,n}v_{p,2,n} \\
m_1(v_{p+1,3,n+1}+v_{p-1,3,n+1})+(1-2m_1)v_{p,3,n+1}+zbu_{p,3,n}-zu^2{}_{p,3,n}v_{p,3,n} \\
m_1(v_{p+1,4,n+1}+v_{p-1,4,n+1})+(1-2m_1)v_{p,4,n+1}+zbu_{p,4,n}-zu^2{}_{p,4,n}v_{p,4,n} \\
. \\ . \\ . \\ . \\
m_1(v_{p+1,M-1,n+1}+v_{p-1,M-1,n+1})+(1-2m_1)v_{p,M-1,n+1}+zbu_{p,M-1,n}-zu^2{}_{p,M-1,n}v_{p,M-1,n}
\end{bmatrix}
$$

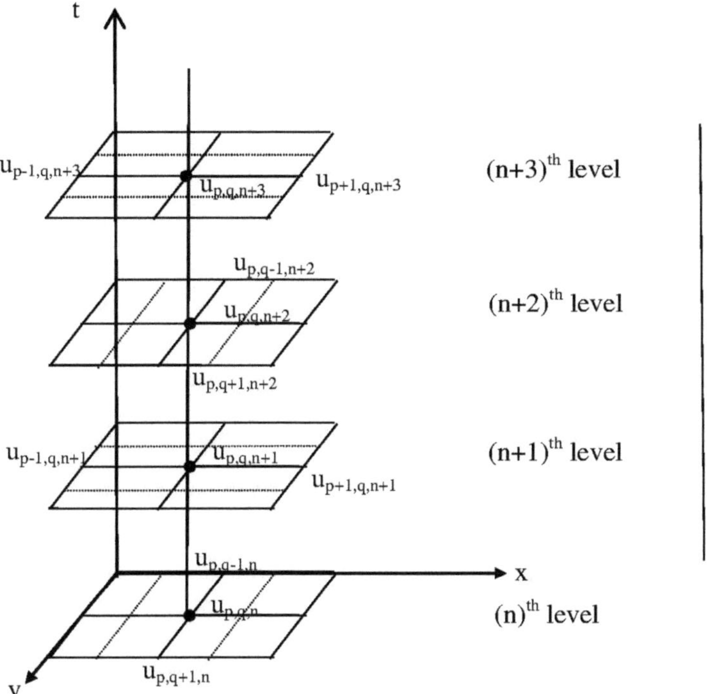

Figure 2.4: The mesh of ADI method

2.4 The Numerical Stability

The numerical stability of the numerical methods is studying the errors introduced by the truncation of the series which are used to represent the derivatives in the process of replacing the differential equation by finite difference equations and the growth of these errors and finding the conditions for which the errors will be decayed from one time step to the next [67].

The Von Neumann analysis is the most commonly used method of determining stability criteria as it is generally the easiest to apply, the most straight-forward and most dependable. This method developed by Von Neumann during World War II, was first discussed in detail by O'Brien, Hyman and Kaplan in a paper published in

1951. The general principle for this method is exchanging the solution of finite difference methods by the value $\xi^q e^{i\beta ph}$ where this obtained as follows:

Assume

$$u_{p,q} = e^{\alpha t} e^{i\beta x} = e^{\alpha q k} e^{i\beta ph} = \xi^q e^{i\beta ph} ,$$

where $\xi = e^{\alpha k}$ and α, in general, is a complex constant and β is a positive real number [77].

2.5 Numerical Stability in One Dimensional Space

Perhaps the most widely used procedure to determine the stability (or instability) of finite difference approximation is the Von Neumann Stability process. In essence, it introduces an initial line of errors as represented by a finite Fourier series and considers the growth (or decay) of these errors as x increases. It is applied only to linear constant coefficient finite difference approximations. If the linearization condition is not met, some form of local linearization is necessary. Because of the linearity, each Fourier component can be treated separately and superposition to add all other components. It is of particular interest that the Von Neumann approach always yields a necessary condition for stability, and in many cases this is also a sufficient condition. Thus, the solution of the finite difference approximation in separable form has the form $u(x,t) = e^{\gamma t} e^{i\beta x} = \psi(t) e^{i\beta x}$ where $i = \sqrt{-1}$ and β is a real number [39].

2.5.1 Numerical Stability Analysis of Explicit Method by Von Neumann Method

The general form of this method is to substitute the solution in finite difference method at the time t by $\psi(t)e^{i\alpha x}$ when α >0 and $i = \sqrt{-1}$ [68].

To apply the Von Neumann on the system(1.1), we go to linearize the system (2.24) [21] :

$$u_{p,q+1} = r_1(u_{p+1,q} + u_{p-1,q}) + (1 - 2r_1 - k(b+1))u_{p,q} + ka$$
$$v_{p,q+1} = r_2(v_{p+1,q} + v_{p-1,q}) + (1 - 2r_2)v_{p,q} + kbu_{p,q},$$

where $r_1 = \frac{d_1 k}{h^2}$ *and* $r_2 = \frac{d_2 k}{h^2}$, $\Delta x = h$ *and* $\Delta t = k$, for the first equation of the system, for some values of *a*, *ka* is zero [46]. So

$$\psi(t+\Delta t)e^{i\alpha x} = r_1[\psi(t)e^{i\alpha(x+\Delta x)} + \psi(t)e^{i\alpha(x-\Delta x)}) + (1-2r_1 - k(b+1)\psi(t)e^{i\alpha x}.$$

Dividing both sides of the above equation by $e^{i\alpha x}$, we obtain:

$$\psi(t+\Delta t) = \psi(t)[r_1(2\cos(\alpha\Delta x)] + \psi(t)[1-2r_1 - k(b+1)]$$

$$\psi(t+\Delta t) = \psi(t)[2r_1(1-2\sin^2(\alpha\Delta x/2)] + \psi(t)[1-2r_1 - k(b+1)].$$

For some values of α, we can assume that $\sin^2(\alpha\Delta x/2)$ is one [67], and

$$\frac{\psi(t+\Delta t)}{\psi(t)} = (1-4r_1 - k(b+1)) = \eta$$

It is stable if $|\frac{\psi(t+\Delta t)}{\psi(t)}| \le 1$, so

$|(1-4r_1 - k(b+1))| \le 1$, which implies $-1 \le 1-4r_1 - k(b+1)) \le 1$

Case 1: $-1 \le 1-4r_1 - k(b+1)) \Rightarrow 4r_1 \le 2-k(b+1) \Rightarrow r_1 \le \frac{2-k(b+1)}{4}$, or

Case 2: $1-4r_1 - k(b+1)) \le 1 \Rightarrow r_1 \ge \frac{-k(b+1)}{4}$ which implies that, $r_1 \ge 0$

The equation is stable under the conditions $r_1 \le \frac{2-k(b+1)}{4}$. For the second equation of the system (1.1):

$$v_{p,q+1} = \frac{d_2 k}{h^2}(v_{p+1,q} - 2v_{p,q} + v_{p-1,q}) + v_{p,q} + kbu_{p,q}.$$

Assuming that $r_2 = \frac{d_2 k}{h^2}$ and for some values of *b*, *kb* is zero, so the equation will be in the form

$$v_{p,q+1} = r_2(v_{p+1,q} + v_{p-1,q}) + (1-2r_2)v_{p,q}.$$

To study the stability of the second equation, we let $v_{p,q} = \varphi(t)e^{i\beta x}$ so the equation will be in the form

$$\varphi(t+\Delta t)e^{i\beta x} = r_2[\varphi(t)e^{i\beta(x+\Delta x)} + \varphi(t)e^{i\beta(x-\Delta x)}] + (1-2r_2)\varphi(t)e^{i\beta x}$$

$$= r_2\varphi(t)[e^{i\beta x}(2\cos(\beta\Delta x/2)) + (1-2r_2)\varphi(t)e^{i\beta x}$$

Dividing both sides of the above equation by $\varphi(t)e^{i\beta x}$, we get :

$$\frac{\phi(t+\Delta t)}{\phi(t)} = r_2(2\cos(\beta\Delta x)) + (1-2r_2)$$
$$= 2r_2(1-2\sin^2(\beta\Delta x/2)) + (1-2r_2)$$
$$= 1-4r_2\sin^2(\beta\Delta x/2) = \eta \ .$$

It is stable if $|\eta| \le 1$, or $\left|\frac{\varphi(t+\Delta t)}{\varphi(t)}\right| \le 1 \Rightarrow |1\text{-}4r_2\sin^2(\beta\Delta x/2)| \le 1$ for some values of β, we can assume that $\sin^2(\beta\Delta x/2)$ is one [67], so

$$|1-4r_2| \le 1 \Rightarrow -1 \le (1-4r_2) \le 1,$$

Case 1: $-1 \le (1-4r_2) \Rightarrow 4r_2 \le 2 \Rightarrow r_2 \le 1/2$

Case 2: $(1-4r_2) \le 1 \Rightarrow 4r_2 \ge 0 \Rightarrow r_2 \ge 0$

And this is always true. Thus the equation is conditionally stable with condition $r_2 \le 1/2$.

Finally the system is conditionally stable under the conditions $r_1 \le \frac{2-k(b+1)}{4}$, and $r_2 \le 1/2$.

2.5.2 Numerical Stability of Implicit Method

We use Crank-Nicolson finite difference in the first equation of the system (1.1) to obtain :

$$\frac{u_{p,q+1}-u_{p,q}}{k} = \frac{d_1}{2h^2}[u_{p-1,q} - 2u_{p,q} + u_{p+1,q} + u_{p-1,q+1} - 2u_{p,q+1} + u_{p+1,q+1}] - (b+1)u_{p,q}.$$

Substituting $u_{p,q}$ by $\psi(t)e^{i\gamma x}$ in the above equation, yields :

$$\frac{\psi(t+\Delta t)e^{i\gamma x} - \psi(t)e^{i\gamma x}}{k} = \frac{d_1}{2h^2}[\psi(t)e^{i\gamma(x-\Delta x)} - 2\psi(t)e^{i\gamma x} + \psi(t)e^{i\gamma(x+\Delta x)} + \psi(t+\Delta t)e^{i\gamma(x-\Delta x)}$$
$$- 2\psi(t+\Delta t)e^{i\gamma x} + \psi(t+\Delta t)e^{i\gamma(x+\Delta x)}] - (b+1)\psi(t)e^{i\gamma x}.$$

Dividing both sides of the above equation by $e^{i\gamma x}$, we obtain :

$$\frac{\psi(t+\Delta t)-\psi(t)}{k} - \frac{d_1}{2h^2}[\psi(t)e^{-i\gamma\Delta x} - 2\psi(t) + \psi(t)e^{i\gamma\Delta x} + \psi(t+\Delta t)e^{-i\gamma\Delta x}$$
$$- 2\psi(t+\Delta t) + \psi(t+\Delta t)e^{i\gamma\Delta x}] = -(b+1)\psi(t).$$

Multiplying both sides of the above equation by k, we get :

$$[\psi(t+\Delta t)-\psi(t)] - \frac{d_1 k\psi(t)}{2h^2}[e^{-i\gamma\Delta x} - 2 + e^{i\gamma\Delta x}] - \frac{d_1 k\psi(t+\Delta t)}{2h^2}[e^{-i\gamma\Delta x} - 2 + e^{i\gamma\Delta x}] = -(b+1)k\psi(t)$$

Asuming $r_1 = \frac{d_1 k}{h^2}$ and

$$[\psi(t+\Delta t)-\psi(t)]-\frac{r_1\psi(t)}{2}[2\cos(\gamma\Delta x)-2]-\frac{r_1\psi(t+\Delta t)}{2}[2\cos(\gamma\Delta x)-2]=-(b+1)k\psi(t)$$

$$[\psi(t+\Delta t)-\psi(t)]+r_1\psi(t)[1-\cos(\gamma\Delta x)]+r_1\psi(t+\Delta t)[1-\cos(\gamma\Delta x)]=-(b+1)k\psi(t)$$

$$[\psi(t+\Delta t)-\psi(t)]+r_1\psi(t)[2\sin^2(\gamma\Delta x/2)]+r_1\psi(t+\Delta t)[2\sin^2(\gamma\Delta x/2)]=-(b+1)k\psi(t),$$

and

$$[1+2r_1\sin^2(\gamma\Delta x/2)]\psi(t+\Delta t)=[1-2r_1\sin^2(\gamma\Delta x/2)-(b+1)k]\psi(t),$$

which implies that

$$\frac{\psi(t+\Delta t)}{\psi(t)}=\frac{1-(2r_1\sin^2(\gamma\Delta x/2)+(b+1)k)}{(1+2r_1\sin^2(\gamma\Delta x/2)}=\xi_i, i=1,2.$$

For stability we need $\left|\frac{\psi(t+\Delta t)}{\psi(t)}\right|\le 1,$ that is.,

$$\left|\frac{1-(2r_1\sin^2(\gamma\Delta x/2)+(b+1)k)}{1+2r_1\sin^2(\gamma\Delta x/2}\right|\le 1, \textit{ for all } r_1,k,b$$

Hence the Crank-Nicolson method is unconditionally stable for the first equation of Brusselator model, and for the second equation

$$\frac{v_{p,q+1}-v_{p,q}}{k}=\frac{d_2}{2h^2}[v_{p-1,q}-2v_{p,q}+v_{p+1,q}+v_{p-1,q+1}-2v_{p,q+1}+v_{p+1,q+1}].$$

Substitute $v_{p,q}=\varphi(t)e^{i\phi x}$ in the above equation to obtain :

$$\frac{\varphi(t+\Delta t)e^{i\phi x}-\varphi(t)e^{i\phi x}}{k}=$$

$$\frac{d_2}{2h^2}[\varphi(t)e^{i\phi(x-\Delta x)}-2\varphi(t)e^{i\phi x}+\varphi(t)e^{i\phi(x+\Delta x)}+\varphi(t+\Delta t)e^{i\phi(x-\Delta x)}$$
$$-2\varphi(t+\Delta t)e^{i\phi x}+\varphi(t+\Delta t)e^{i\phi(x+\Delta x)}].$$

Multiplying both sides of the above equation by $k\ e^{-i\beta x}$, we obtain :

$$\varphi(t+\Delta t)-\varphi(t)=\frac{d_2 k\varphi(t)}{2h^2}[e^{-i\phi\Delta x}-2+e^{i\phi\Delta x}]+\frac{d_2 k\varphi(t+\Delta t)}{2h^2}[e^{-i\phi\Delta x}-2+e^{i\phi\Delta x}].$$

Let $r_2=\frac{d_2 k}{h^2}\Rightarrow \varphi(t+\Delta t)-\varphi(t)=\frac{r_2\varphi(t+\Delta t)}{2}[2\cos(\phi\Delta x)-2]+r_2\frac{\varphi(t)}{2}[2\cos(\phi\Delta x)-2],$

$$\varphi(t+\Delta t)-\varphi(t)=-r_2\varphi(t+\Delta t)[1-\cos(\phi\Delta x)]-r_2\varphi(t)[1-\cos(\phi\Delta x)],$$

$$\varphi(t+\Delta t)-\varphi(t)=-r_2\varphi(t+\Delta t)[1-(1-2\sin^2(\phi\Delta x/2)]-r_2\varphi(t)[1-(1-2\sin^2(\phi\Delta x/2)]$$

$$\varphi(t+\Delta t)-\varphi(t) = -2r_2\varphi(t+\Delta t)\sin^2(\phi\Delta x/2)-2r_2\varphi(t)\sin^2(\phi\Delta x/2)$$

$$\varphi(t+\Delta t)+2r_2\varphi(t+\Delta t)\sin^2(\phi\Delta x/2) = \varphi(t) -2r_2\varphi(t)\sin^2(\phi\Delta x/2)$$

$$[1+2r_2 \sin^2(\phi\Delta x/2)]\,\varphi(t+\Delta t)=[1-2r_2 \sin^2(\phi\Delta x/2)]\,\varphi(t)$$

$$\frac{\phi(t+\Delta t)}{\phi(t)} = \frac{[1\text{-}2\mathrm{r}_2\sin^2(\varphi\Delta \mathrm{x}/2)\,]}{[\,1+2\mathrm{r}_2\sin^2\,(\varphi\Delta \mathrm{x}/2)]} = \xi_i\,, i=1,2.$$

Thus $|\frac{\varphi(t+\Delta t)}{\varphi(t)}|=|\xi|\le 1$ so

$$|\frac{[1\text{-}2\mathrm{r}_2\sin^2(\phi\Delta \mathrm{x}/2)\,]}{[1+2\mathrm{r}_2\sin^2\,(\phi\Delta \mathrm{x}/2)]}|\le 1,\ \forall r_2, b, k,$$

Since for both equations of the system we have $|\ \xi_i\ |\le 1$, the Crank-Nicolson method is unconditionally stable.

2.6 **Numerical Stability in Two Dimensional Space**

2.6.1 Numerical stability of ADE

The Von-Neumann method is used to study the stability analysis of Brusselator Model in two dimensions. We can apply this method by substituting the solution in finite difference method at time t by $\psi(t)e^{i\beta x}e^{i\gamma y}$ where β, and $\gamma>0$, to apply Von-Neumann on the first equation of Brusselator Model :

$$\frac{\partial u}{\partial t} = d_1[\frac{\partial^2 u}{\partial x^2}+\frac{\partial^2 u}{\partial y^2}]-(b+1)u+a,$$

and after eliminating the nonlinear term, then the above equation becomes

$$u_{p,q,n+1} = (1-4r_1-k(b+1))u_{p,q,n}+r_1(u_{p+1,q,n}+u_{p-1,q,n}+u_{p,q+1,n}+u_{p,q-1,n})+ak$$

Neglecting ak for some values of a in [2], where $r_1 = \frac{d_1 k}{h^2}$, so the equation will be in the form

$$\psi(t+\Delta t)e^{m\beta x}e^{m\gamma y} = (1-4r_1-k(b+1))\psi(t)e^{m\beta x}e^{m\gamma y} + r_1(\psi(t)e^{m\beta(x+\Delta x)}e^{m\gamma y} + \psi(t)e^{m\beta(x-\Delta x)}e^{m\gamma y} + \psi(t)e^{m\beta x}e^{m\gamma(y+\Delta y)} + \psi(t)e^{m\beta x}e^{m\gamma(y-\Delta y)}.$$

Now dividing both sides of the above equation by $\psi(t)e^{i\beta x}e^{i\gamma y}$, we obtain :

$$\frac{\psi(t+\Delta t)}{\psi(t)} = (1-4r_1 - k(b+1)) + r_1(e^{m\beta\Delta x} + e^{-m\beta\Delta x} + e^{m\beta\Delta y} + e^{-m\beta\Delta y})$$
$$= (1-4r_1 - k(b+1)) + r_1(2\cos(\beta\Delta x) + 2\cos(\gamma\Delta y))$$
$$= (1-4r_1 - k(b+1)) + r_1(1-2\sin^2(\beta\Delta x/2) + 1 - 2\sin^2(\gamma\Delta y/2)).$$

For some values of β and γ, $\sin^2(\beta\Delta x/2)$ and $\sin^2(\gamma\Delta y/2)$ is unity [67], so

$$\frac{\psi(t+\Delta t)}{\psi(t)} = (1-4r_1 - k(b+1)) + 2r_1(-2) = (1-8r_1 - k(b+1)) = \xi .$$

For stable situation, we need $|\xi| \le 1$, so $-1 \le (1-8r_1 - k(b+1)) \le 1$,

Case 1: $-1 \le (1-8r_1 - k(b+1) \Rightarrow r_1 \le \frac{2-k(b+1)}{8}$

Case 2: $(1-8r_1 - k(b+1)) \le 1 \Rightarrow r_1 \ge \frac{-k(b+1)}{8} \Rightarrow r_1 \ge 0$.

For the second (Linearized) equation of Brusselator model which is in the form :

$\frac{\partial v}{\partial t} = d_2(\frac{\partial^2 v}{\partial x^2} + \frac{\partial^2 v}{\partial y^2}) + bu$, for small values of "b" we can neglect bu, so we have

$\frac{\partial v}{\partial t} = d_2(\frac{\partial^2 v}{\partial x^2} + \frac{\partial^2 v}{\partial y^2})$ and we apply the ADE method on this equation, to have :

$$\frac{v_{p,q,n+1} - v_{p,q,n}}{z} = \frac{d_2}{h^2}[v_{p+1,q,n} - 2v_{p,q,n} + v_{p-1,q,n} + v_{p,q+1,n} - 2v_{p,q,n} + v_{p,q-1,n}]$$

Assuming that $\Delta x = \Delta y = h,\ and\ r_2 = \frac{d_2 k}{h^2}$

$$v_{p,q,n+1} = (1-4r_2)v_{p,q,n} + r_2(v_{p+1,q,n} + v_{p-1,q,n} + v_{p,q+1,n} + v_{p,q-1,n}) .$$

To study the stability by Von-Neumann, let $V_{p,q,n} = \varphi(t)e^{m\beta x}e^{m\gamma y}$ be substituted in the above equation to obtain

$$\varphi(t+\Delta t)e^{m\beta x}e^{m\gamma y} = (1-4r_2) + r_2\varphi(t)e^{m\beta(x+\Delta x)}e^{m\gamma y} + r_2\varphi(t)e^{m\beta(x-\Delta x)}e^{m\gamma y} + r_2\varphi(t)e^{m\beta x}e^{m\gamma(y+\Delta y)} +$$
$$+ r_2\varphi(t)e^{m\beta x}e^{m\gamma(y-\Delta y)}$$

Dividing both sides of the above equation by $\varphi(t)e^{m\beta x}e^{m\gamma y}$, we obtain :

$$\frac{\phi(t+\Delta t)}{\phi(t)} = \xi = (1-4r_2) + m_2[e^{m\beta\Delta x} + e^{-m\beta\Delta x} + e^{m\gamma\Delta y} + e^{-m\gamma\Delta y}]$$
$$= (1-4m_2) + r_2[2\cos(\beta\Delta x) + 2\cos(\gamma\Delta y)]$$
$$= (1-4r_2) + 2r_2[1-2\sin^2(\beta\Delta x/2) + 1 - 2\sin^2(\gamma\Delta y/2)]$$
$$= (1-4r_2\sin^2(\beta\Delta x/2) - 4r_2\sin^2(\gamma\Delta y/2).$$
$$= (1-4r_2[\sin^2(\beta\Delta x/2) + \sin^2(\gamma\Delta y/2)])$$

For some values of β and γ, we can assume that $\sin^2(\beta\Delta x/2)$ and $\sin^2(\gamma\Delta y/2)$ are unity [67], so

$\frac{\varphi(t+\Delta t)}{\varphi(t)} = \xi = (1-8r_2)$. The equation is stable if $|\xi| \le 1$ which implies that

$|1-8r_2| \le 1 \Rightarrow -1 < (1-8r_2) < 1$, which are located in two cases

Case 1: $-1 < (1-8r_2) \Rightarrow 8r_2 < 2 \Rightarrow r_2 < 1/4$ and

Case 2: $(1-8r_2) < 1 \Rightarrow 8r_2 \ge 0 \Rightarrow r_2 \ge 0$.

So the system is stable under the conditions $r_1 \le \frac{2-k(b+1)}{8}$,and $r_2 < 1/4$.

2.6.2 Numerical Stability of ADI

The ADI finite difference form for the first equation of (1.1) is :

$$(1+2r_1)u_{p,q,n+1} = r_1(u_{p+1,q,n+1} + u_{p-1,q,n+1}) + r_1(u_{p,q+1,n} + u_{p,q-1,n}) +$$
$$(1-2r_1 - k(b+1))u_{p,q,n} .$$

In order to study the stability of the above equation, let $u_{p,q,n} = \psi(t)e^{m\beta x}e^{m\gamma y}$, where $m = \sqrt{-1}$,

$$(1+2r_1)\psi(t+\Delta t)e^{m\beta x}e^{m\gamma y} = r_1(\psi(t+\Delta t)e^{m\beta(x+\Delta x)}e^{m\gamma y} + \psi(t+\Delta t)e^{m\beta(x-\Delta x)}e^{m\gamma y}) +$$
$$r_1(\psi(t)e^{m\beta x}e^{m\gamma(y+\Delta y)} + \psi(t)e^{m\beta x}e^{m\gamma(y-\Delta y)}) + (1-2r_1-k(b+1))\psi(t)e^{m\beta x}e^{m\gamma y} .$$

Dividing both sides of the above equation by $e^{m\beta x}e^{m\gamma y}$,we get :

$$(1+2r_1)\psi(t+\Delta t) = r_1(\psi(t+\Delta t)e^{m\beta\Delta x} + \psi(t+\Delta t)e^{-m\beta\Delta x}) + r_1(\psi(t)e^{m\gamma\Delta y} +$$
$$\psi(t)e^{-m\gamma\Delta y} + (1-2r_1-k(b+1))\psi(t) .$$

Rearranging the above equation, we have :

$$(1+2r_1-r_1e^{m\beta\Delta x}-r_1e^{-m\beta\Delta x})\psi(t+\Delta t)=r_1\psi(t)(e^{m\gamma\Delta y}+e^{-m\gamma\Delta y})+$$
$$(1-2r_1-k(b+1))\psi(t)\Rightarrow(1+2r_1-r_1(2\cos(\beta\Delta x))\psi(t+\Delta t)=r_1\psi(t)(2\cos(\gamma\Delta t))+$$
$$(1-2r_1-k(b+1))\psi(t)\Rightarrow(1+2r_1-r_1(2(1-2\sin^2(\beta\Delta x/2))\psi(t+\Delta t)=$$
$$r_1\psi(t)(2(1-2\sin^2(\gamma\Delta y/2))+(1-2r_1-k(b+1))\psi(t).$$

For some values of β and γ, assume that $\sin^2(\beta\Delta x/2)$ and $\sin^2(\gamma\Delta y/2)$ are unity [67] , so the equation will take the form :

$$\Rightarrow(1+2r_1-2r_1(-1)\psi(t+\Delta t)=r_1\psi(t)(2(-1))+$$
$$(1-2r_1-k(b+1))\psi(t).$$
$$\Rightarrow(1+4r_1)\psi(t+\Delta t)=-2r_1\psi(t)+(1-2r_1-k(b+1))\psi(t)$$
$$\Rightarrow\frac{\psi(t+\Delta t)}{\psi(t)}=\frac{-2r_1+(1-2r_1-k(b+1)}{1+4r_1}=\xi_1 \ .$$

So $\dfrac{\psi(t+\Delta t)}{\psi(t)}=\dfrac{1-(4r_1+k(b+1))}{1+4r_1}=\xi_1$.

Similarly for the second equation of the Brusselator model we assume that $v_{p,q,n}=\varphi(t)e^{m\beta x}e^{m\gamma y}$, where $m=\sqrt{-1}$,the Finite difference form of this euation is

$$\frac{v_{p,q,n+1}-v_{p,q,n}}{z}=\frac{d_2}{h^2}[(v_{p+1,q,n+1}-2v_{p,q.n+1}+v_{p-1,q,n+1})+\frac{d_2}{k^2}(v_{p,q+1,n}-2v_{p,q,n}+v_{p,q-1,n})$$

$$where\ \ m_1=\frac{d_2z}{h^2},\ \ and\ \ m_2=\frac{d_2z}{k^2},\ \ if\ \ h=k,m_1=m_2=m$$

$$v_{p,q,n+1}=m(v_{p+1,q,n+1}-2v_{p,q.n+1}+v_{p-1,q,n+1})+m(v_{p,q+1,n}-2v_{p,q,n}+v_{p,q-1,n})$$
$$(1+2m)v_{p,q,n+1}=m(v_{p+1,q,n+1}+v_{p-1,q,n+1})+(1-2m)v_{p,q,n}+m(v_{p,q+1,n}+v_{p,q-1,n}).$$

Substituting $v_{p,q,n}=\varphi(t)e^{m\beta x}e^{m\gamma y}$, where $m=\sqrt{-1}$ in the above equation to get

$$(1+2m)\phi(t+\Delta t)e^{m\beta x}e^{m\gamma y}-m(\phi(t+\Delta t)e^{m\beta(x+\Delta x)}e^{m\gamma y}+\phi(t+\Delta t)e^{m\beta(x-\Delta x)}e^{m\gamma y})=$$
$$(1-2m)\phi(t)e^{m\beta x}e^{m\gamma y}+m\phi(t)(e^{m\beta x}e^{m\gamma(y+\Delta y)}+e^{m\beta x}e^{m\gamma(y-\Delta y)})+\phi(t)e^{m\beta x}e^{m\gamma y}$$

Dividing both sides of the equation by $e^{m\beta x}e^{m\gamma y}$, to get

$$[(1+2m)-m(e^{m\beta\Delta x}+e^{-m\beta\Delta x})]\phi(t+\Delta t)=m[e^{m\gamma\Delta y}+e^{-m\gamma\Delta y}]\phi(t)+(1-2m)\phi(t)e^{m\beta x}e^{m\gamma y}$$

$$[(1+2m)-m(2\cos(\beta\Delta x)]\phi(t+\Delta t)=m[2\cos(\gamma\Delta y)]\phi(t)+(1-2m)\phi(t)$$

Which implies that

$$[(1+2m)-2m(\cos(\beta\Delta x)]\phi(t+\Delta t)=2m[1-\cos(\gamma\Delta y)]\phi(t)+\phi(t)$$

$$\Rightarrow [1+4m\sin^2(\beta\Delta x/2)]\phi(t+\Delta t)=-4m\sin^2(\gamma\Delta y/2)\phi(t)+\phi(t)$$

we have $(1+4m)\phi(t+\Delta t)=(1-4m)\phi(t)$ and $\xi_2=\dfrac{\phi(t+\Delta t)}{\phi(t)}=\dfrac{1-4m}{1+4m}$ where ξ_1 *and* ξ_2 stand for the I-plane and II-plane respectively, each of the above terms ξ_1 *and* ξ_2 are conditionally stable. However the combined two-level has the form

$$\xi_{ADI}=\xi_1.\xi_2\left[\frac{1-(4r_1+z(b+1))}{1+4r_1}\right].\left[\frac{1-4m}{1+4m}\right].$$

Thus the above scheme is unconditionally stable, each individual equation is conditionally stable by itself, and the combined two-level is completely stable.

Thus, the above scheme is unconditionally stable, each individual equation is conditionally stable by itself, and the combined two-level is completely stable.

We conclude that the explicit method for solving Brusselator system is stable under the condition $r_1\le\dfrac{2-k(b+1)}{4}$, and $r_2\le 1/2$.

While the implicit method is unconditionally stable. For two dimension in space, we find that ADE method is stable under condition $r_1\le\dfrac{2-z(b+1)}{8}$, and $r_2<1/4$, while ADI is unconditionally stable.

CHAPTER THREE

Finite Element Method

3.1 Introduction

The finite element method is a numerical analysis technique for obtaining approximate solutions to a wide variety of engineering problem.

Although originally developed to study stresses in complex airframe structures, it has since been extended and applied to the broad field of continuum mechanics. Because of its diversity and flexibility as an analysis tool, it has received much attention in engineering schools and in industry. This brief comment on the finite element method answers the question posed by the section heading. It does not give us the operational definition, so we need to apply the method to a particular problem. Such an operational definition-along with a description of the fundamentals of the method requires, considerably, more than one paragraph to develop. In more and more engineering situations today, we find that it is necessary to obtain approximate numerical solutions to problems rather than exact closed-form solutions. The finite element method has been developed simultaneously with the increasing use of high-speed electronic digital computers and with the growing emphasis on numerical methods for engineering analysis. Although the method was originally developed for structural analysis, the general nature of the theory on which it was based has also made possible its successful application for solutions of problems in other fields of engineering [17].

In this chapter, we study the numerical solution for Brusselator equations (1.1) by finite element method .

3.2 Numerical Solutions

Galerkin method has been discussed by several authors (see [31] , [15]) for obtaining an approximate solution to differential equation. It is carried out by requiring that the error between the approximate solution and the exact solution be orthogonal to the function used in the approximation. If we start a differential equation $lu-f=0$ (l is a differential operator) and approximate the solution by $u^*=\sum N_i u_i$, then the solution $lu^*-f=\zeta$ where ζ is a residual or error. Our desire is to make ζ as small as possible. One way of accomplishing this objective is to require the integral $\int_R N_i \zeta dx=0$ for each basis function N_i. This integral mathematically states the basis function which must be orthogonal to the error over the region [58]. Let u^* have the normal finite element form

$$u \approx u^* = \sum_{e=1}^{E} u^{(e)}, \tag{3.1}$$

where the interpolation function is [3]:

$$\left.\begin{aligned} u^{(e)} &= N_i^{(e)} u_i + N_j^{(e)} u_j \\ v^{(e)} &= N_i^{(e)} v_i + N_j^{(e)} v_j \end{aligned}\right\} \tag{3.2}$$

where $N_i^{(e)}$ and $N_j^{(e)}$ are pyramid functions :

$$N_j^{(e)} = \frac{1}{L}(a_j + b_j x) \quad \text{such that} \quad a_j = -x_i \quad \text{and} \quad b_j = 1$$

$$N_i^{(e)} = \frac{1}{L}(a_i + b_i x) \quad \text{such that} \quad a_i = x_j \quad \text{and} \quad b_i = -1 \tag{3.3}$$

where L represents interval length (step size) for x, and the derivative for equation (3.2) is:

$$\left.\begin{aligned} &\frac{du^{(e)}}{dx} = \frac{dN_i}{dx} u_i + \frac{dN_j}{dx} u_j, \\ \text{and} \\ &\frac{dv^{(e)}}{dx} = \frac{dN_i}{dx} v_i + \frac{dN_j}{dx} v_j. \end{aligned}\right\} \tag{3.4}$$

Also, integral over the element must involve the integrals of the pyramid functions [3]:

$$\int_L N_i dx = \int_L N_j dx = \frac{1}{2} L. \tag{3.5}$$

Thus,

$$\left.\begin{aligned}\int_L u^{(e)}dx &= \frac{1}{2}L(u_i + u_j), \\ \int_L v^{(e)}dx &= \frac{1}{2}L(v_i + v_j)\end{aligned}\right\} \tag{3.6}$$

It has been shown that the general integration formula is [3] :

$$\int_L N_i^{\alpha} N_j^{\beta} dx = \frac{\alpha!\beta!}{(\alpha+\beta+1)!}L \tag{3.7}$$

Galerkin's method uses the approximating function as weighting functions:

$$W_n = N_j^{(e)} + N_i^{(e+1)} \tag{3.8}$$

3.2.1 Element Residuals

The element residuals R_i, R_j associated with the nodes *i, j* are [3]:

$$R_i^{(e)} = -\int_{L^{(e)}} W_i^{(e)} r^{(e)} dx \ ,$$

$$R_j^{(e)} = -\int_{L^{(e)}} W_j^{(e)} r^{(e)} dx \ ,$$

where $R_i^{(e)}$ is contribution of element (*e*) to residual in the first equation of (1.1) for node (*i*), and $R_j^{(e)}$ is contribution of element (*e*) to residual in the first equation of (1.1) for node (*j*).

We will consider a family of semi-implicit methods called "*one step theta schemes*". These depend on a parameter θ, $0 \le \theta \le 1$, and include the backward difference approximation (θ=0), the Crank-Nicolson (θ=1/2), and the forward difference approximation (θ=1) [76] .

For the second equation of (1.1), the element residuals O_i, O_j associated with the nodes i, j are :

$$O_i^{(e)} = -\int_{L^{(e)}} W_i^{(e)} o^{(e)} dx$$

$$O_j^{(e)} = -\int_{L^{(e)}} W_j^{(e)} o^{(e)} dx$$

Galerkin's method [31] is employed for the finite element formulation in space and time. After substituting the approximation $u^*(x,t)$ and $v^*(x,t)$, the errors for the equation (1.1) is :

$$r(x,t)=\sum_{e=1}^{E} r^{(e)}=\sum_{e=1}^{E}[d_1((1-\theta)\frac{\partial^2 u^m}{\partial x^2}+\theta\frac{\partial^2 u^{m+1}}{\partial x^2})-(b+1)u^m+(u^m)^2 v^m-(\frac{u^{m+1}-u^m}{\Delta t})+a] \quad (3.9)$$

$$O(x,t)=\sum_{e=1}^{E} O^{(e)}=\sum_{e=1}^{E}[d_2((1-\theta)\frac{\partial^2 v^m}{\partial x^2}+\theta\frac{\partial^2 v^{m+1}}{\partial x^2})+bu^m-(u^m)^2 v^m-(\frac{v^{m+1}-v^m}{\Delta t})] \quad (3.10)$$

Where E is the total number of elements.

For the (i^{th}) node in element (e) in equation (3.9), the residual is:

$R_i^e=[R_i]_{space}^{(e)}+[R_i]_{time}^{(e)}$

where

$$[R_i]_{space}^{(e)}=-\int_{(e)}[N_i[d_1\theta\frac{\partial^2 u^{m+1}}{\partial x^2}+d_1(1-\theta)\frac{\partial^2 u^m}{\partial x^2}-(b+1)u^m+(u^m)^2 v^m+a]]^{(e)}dx \quad (3.11)$$

$$[R_i]_{time}^{(e)}=-\int_{L^{(e)}}\left(N_i\left[-\left(\frac{u^{m+1}-u^m}{\Delta t}\right)\right]\right)^{(e)}dx, \quad (3.12)$$

also for the (j^{th}) node in element (e) in equation (3.9), the residual is:

$$R_j^{(e)}=[R_j]_{space}^{(e)}+[R_j]_{time}^{(e)}, \quad (3.13)$$

where

$$[R_j]_{space}^{(e)}=-\int_{(e)}[N_j[d_1\theta\frac{\partial^2 u^{m+1}}{\partial x^2}+d_1(1-\theta)\frac{\partial^2 u^m}{\partial x^2}-(b+1)u^m+(u^m)^2 v^m+a]]^{(e)}dx \quad (3.14)$$

and

$$[R_j]_{time}^{(e)}=-\int_{L^{(e)}}\left(N_j\left[-\left(\frac{u^{m+1}-u^m}{\Delta t}\right)\right]\right)^{(e)}dx. \quad (3.15)$$

For the (i^{th}) node in element (e) in equation (3.10), the residual is:

$O_i^{(e)}=[O_i]_{space}^{(e)}+[O_i]_{time}^{(e)}$

$$[O_i]_{space}^{(e)}=-\int_{(e)}[N_i[d_2\theta\frac{\partial^2 v^{m+1}}{\partial x^2}+d_2(1-\theta)\frac{\partial^2 v^m}{\partial x^2}+bu^m-(u^m)^2 v^m]]^{(e)}dx\ . \quad (3.16)$$

$$[O_i]_{time}^{(e)}=-\int_{L^{(e)}}\left(N_i\left[-\left(\frac{v^{m+1}-v^m}{\Delta t}\right)\right]\right)^{(e)}dx\ . \quad (3.17)$$

And for the (j^{th}) node in element (e) in the equation (3.10), the residual is:

$O_j^{(e)}=[O_j]_{space}^{(e)}+[O_j]_{time}^{(e)}$.

$$[O_j]_{space}^{(e)} = -\int_{(e)} [N_j[d_2\theta \frac{\partial^2 v^{m+1}}{\partial x^2} + d_2(1-\theta)\frac{\partial^2 v^m}{\partial x^2} + bu^m - (u^m)^2 v^m]]^{(e)} dx. \quad (3.18)$$

$$[O_j]_{time}^{(e)} = -\int_{L^{(e)}} \left(N_j \left[-\left(\frac{v^{m+1} - v^m}{\Delta t} \right) \right] \right)^{(e)} dx . \quad (3.19)$$

Now, we use the integration by parts and use the Neumann boundary conditions (1.3), and from (3.2) , (3.4) , (3.5) and (3.7), the equation (3.11) becomes

$$[R_i]_{space}^{(e)} = -\begin{pmatrix} d_1\theta(-\frac{b_i b_i}{L} u_i^{m+1} - \frac{b_i b_j}{L} u_j^{m+1}) + d_1(1-\theta)(-\frac{b_i b_i}{L} u_i^m - \frac{b_i b_j}{L} u_j^m) - (b+1)\frac{L}{3} u_i^m - (b+1)\frac{L}{6} u_j^m \\ + \frac{L}{5}(u_i^m)^2 v_i^m + \frac{L}{10} u_i^m u_j^m v_i^m + \frac{L}{30}(u_j^m)^2 v_i^m + \frac{L}{20}(u_i^m)^2 v_j^m + \frac{L}{15} u_i^m u_j^m v_j^m + \frac{L}{20}(u_j^m)^2 v_j^m + \frac{aL}{2} \end{pmatrix} \quad (3.20)$$

From the time term (4.12) in the (i^{th}) node, we get

$$[R_i]_{time}^{(e)} = \int_{L^{(e)}} N_i(\frac{u^{m+1} - u^m}{\Delta t})dx = \int_{L^{(e)}} N_i(\frac{N_i u_i^{m+1} + N_j u_j^{m+1}}{\Delta t} - \frac{N_i u_i^m + N_j u_j^m}{\Delta t})dx$$

$$= \frac{1}{\Delta t} \int_{L^{(e)}} (N_i^2 u_i^{m+1} + N_i N_j u_j^{m+1} - N_i^2 u_i^m - N_i N_j u_j^m)dx = \frac{1}{\Delta t}[\frac{L}{3} u_i^{m+1} + \frac{L}{6} u_j^{m+1} - \frac{L}{3} u_i^m - \frac{L}{6} u_j^m] \quad (3.21)$$

Now we will evaluate $[R_j]_{space}^{(e)}$ for the equation (3.11), using the method of one-step theta schemes, gives

$$[R_j]_{space}^{(e)} = -\int_{L^{(e)}} [N_j \left(d_1\theta \frac{\partial^2 u^{m+1}}{\partial x^2} + d_1(1-\theta)\frac{\partial^2 u^m}{\partial x^2} - (b+1)u^m + (u^m)^2(v^m) + a \right)]dx$$

$$= [-d_1\theta \int_{L^{(e)}} N_j \frac{\partial^2 u^{m+1}}{\partial x^2} - d_1(1-\theta) \int_{L^{(e)}} N_j \frac{\partial^2 u^m}{\partial x^2} + (b+1) \int_{L^{(e)}} N_j u^m - \int_{L^{(e)}} N_j (u^m)^2 (v^m) - \int_{L^{(e)}} N_j a]dx. \quad (3.22)$$

and from (3.2) , (3.4) , (3.5) and (3.7), the equation (3.22) becomes

$$[R_j]_{space}^{(e)} = -$$

$$\left(\begin{array}{l} d_1\theta(-b_j \frac{b_i}{L} u_i^{m+1} - b_j \frac{b_j}{L} u_j^{m+1}) + d_1(1-\theta)(-b_j \frac{b_i}{L} u_i^m - b_j \frac{b_j}{L} u_j^m) \\ -(b+1)(\frac{L}{6} u_i^m + \frac{L}{3} u_j^m) + \frac{L}{20}(u_i^m)^2 v_i^m + \frac{L}{15} u_i^m u_j^m v_i^m + \frac{L}{20}(u_j^m)^2 v_i^m + \frac{L}{30}(u_i^m)^2 v_j^m + \\ + \frac{L}{10} u_i^m u_j^m v_j^m + \frac{L}{5}(u_j^m)^2 v_j^m + \frac{aL}{2}. \end{array} \right) \tag{3.23}$$

$$[R_j]_{space}^{(e)} = [\frac{d_1\theta}{L} b_i b_j u_i^{m+1}]^{(e)} + [\frac{d_1\theta}{L} b_j b_j u_j^{m+1}]^{(e)} + [\frac{d_1(1-\theta)}{L} b_i b_j u_i^m]^{(e)} + [\frac{d_1(1-\theta)}{L} b_j b_j u_j^m]^{(e)}$$

$$+[(B+1)\frac{L}{6} u_i^m]^{(e)} + [(b+1)\frac{L}{3} u_j^m]^{(e)} - [\frac{L}{20}(u_i^m)^2 v_i^m]^{(e)} - [\frac{L}{15} u_i^m u_j^m v_i^m]^{(e)} - [\frac{L}{20}(u_j^m)^2 v_i^m]^{(e)}$$

$$-[\frac{L}{30}(u_i^m)^2 v_j^m]^{(e)} - [\frac{L}{10} u_i^m u_j^m v_j^m]^{(e)} - [\frac{L}{5}(u_j^m)^2 v_j^m]^{(e)} - [\frac{aL}{2}]^{(e)}.$$

$$[R_j]_{time}^{(e)} = \frac{1}{\Delta t}[\frac{L}{3} u_i^{m+1} + \frac{L}{6} u_j^{m+1} - \frac{L}{3} u_i^m - \frac{L}{6} u_j^m]. \tag{3.24}$$

And from equation (3.16) ,and using the equations(3.2),(3.4),(3.5),and (3.7) we have

$$[O_i]_{space}^{(e)} = -\int_{(e)} [N_i [d_2\theta \frac{\partial^2 v^{m+1}}{\partial x^2} + d_2(1-\theta)\frac{\partial^2 v^m}{\partial x^2} + bu^m - (u^m)^2 v^m]]^{(e)} dx.$$

$$= -\left(\begin{array}{l} d_2\theta(-\frac{b_i b_i}{L} v_i^{m+1} - \frac{b_i b_j}{L} v_j^{m+1}) + d_2(1-\theta)(-\frac{b_i b_i}{L} v_i^m - \frac{b_i b_j}{L} v_j^m) + \frac{bL}{3} u_i^{m+1} + \frac{bL}{6} u_j^{m+1} - \frac{L}{5}(u_i^{m+1})^2 v_i^m \\ -\frac{L}{10} u_i^{m+1} u_j^{m+1} v_i^m - \frac{L}{30}(u_j^{m+1})^2 v_i^m - \frac{L}{20}(u_i^{m+1})^2 v_j^m - \frac{L}{15} u_i^{m+1} u_j^{m+1} v_j^m - \frac{L}{20}(u_j^{m+1})^2 v_j^m \end{array} \right) \tag{3.25}$$

$$[O_i]_{time}^{(e)} = -\int_{L^{(e)}} \left(N_i \left[-\left(\frac{u^{m+1} - u^m}{\Delta t} \right) \right] \right)^{(e)} dx.$$

$$[O_i]_{time}^{(e)} = \frac{1}{\Delta t}[\frac{L}{3} v_i^{m+1} + \frac{L}{6} v_j^{m+1} - \frac{L}{3} v_i^m - \frac{L}{6} v_j^m]. \tag{3.26}$$

And for the (jth) node in element (e), the equation (3.18) becomes, and using (3.2),(3.4),(3.5), and (3.7) we have :

$$[O_j]_{space}^{(e)} = -\int_{(e)} [N_j [d_2\theta \frac{\partial^2 v^{m+1}}{\partial x^2} + d_2(1-\theta)\frac{\partial^2 v^m}{\partial x^2} + bu^m - (u^m)^2 v^m]]^{(e)} dx.$$

$$[O_j]_{space}^{(e)} = -\left(\begin{array}{l} d_2\theta(-\frac{b_j b_i}{L} v_i^{m+1} - \frac{b_j b_j}{L} v_j^{m+1}) + d_2(1-\theta)(-\frac{b_j b_i}{L} v_i^m - \frac{b_j b_j}{L} v_j^m) + \frac{bL}{6} u_i^{m+1} + \frac{bL}{3} u_j^{m+1} \\ -\frac{L}{20}(u_i^{m+1})^2 v_i^m - \frac{L}{30}(u_i^{m+1})^2 v_j^m - \frac{L}{15} u_i^{m+1} u_j^{m+1} v_i^m - \frac{L}{10} u_i^{m+1} u_j^{m+1} v_j^m - \frac{L}{20}(u_j^{m+1})^2 v_i^m - \frac{L}{5}(u_j^{m+1})^2 v_j^m \end{array} \right).$$

By the same way as in residual (i[th]), we get:

$$[O_j]_{time}^{(e)} = -\int_{L^{(e)}} \left(N_j \left[-\left(\frac{v^{m+1} - v^m}{\Delta t} \right) \right] \right)^{(e)} dx = \frac{1}{\Delta t}[\frac{L}{6} v_i^{m+1} + \frac{L}{3} v_j^{m+1} - \frac{L}{6} v_i^m - \frac{L}{3} v_j^m].$$

3.2.2 Element Matrices

It is convenient to organize the residuals in element matrix form because a set of simultaneous nodal equations must be solved at each time step. Then the normal assembly procedure can be used. The terms involving nodal values at time step (*m+1*) are unknown and appear in the element matrix. All other terms are known and go in the element column [3]. $[R_j]_{space}^{(e)}$, and $[R_j]_{time}^{(e)}$, in equations (3.23) and (3.24) they have the form :

$$\begin{bmatrix} R_i \\ R_j \end{bmatrix}^{(e)} = [B]^{(e)} \begin{bmatrix} u_i^{m+1} \\ u_j^{m+1} \end{bmatrix} - [C]^{(e)} \begin{bmatrix} u_i^m \\ u_j^m \end{bmatrix}$$

the particular components can now be evaluated.

The element residuals are written as:

$$\begin{bmatrix} R_i \\ R_j \end{bmatrix}^{(e)} = \begin{bmatrix} R_i \\ R_j \end{bmatrix}_{space}^{(e)} + \begin{bmatrix} R_i \\ R_j \end{bmatrix}_{time}^{(e)}$$

Substituting the equation of $[R_i]_{space}^{(e)}$ and $[R_j]_{space}^{(e)}$ for the space residual yields:

$$\begin{bmatrix} R_i \\ R_j \end{bmatrix}_{space}^{(e)} = d_1\theta \begin{bmatrix} \frac{b_i b_i}{L} & \frac{b_i b_j}{L} \\ \frac{b_i b_j}{L} & \frac{b_j b_j}{L} \end{bmatrix} \begin{bmatrix} u_i^{m+1} \\ u_j^{m+1} \end{bmatrix} + (1-\theta)d_1 \begin{bmatrix} \frac{b_i b_i}{L} & \frac{b_i b_j}{L} \\ \frac{b_i b_j}{L} & \frac{b_j b_j}{L} \end{bmatrix} \begin{bmatrix} u_i^m \\ u_j^m \end{bmatrix} +$$

$$\begin{bmatrix} \frac{(b+1)L}{3} - \frac{L}{5}u_i^m v_i^m - \frac{L}{10}u_i^m v_j^m - \frac{L}{15}u_j^m v_j^m - \frac{L}{20}u_i^m v_j^m, & \frac{(b+1)L}{6} - \frac{L}{30}u_j^m v_i^m - \frac{L}{15}u_i^m v_j^m - \frac{L}{20}u_j^m v_j^m \\ \frac{(b+1)L}{6} - \frac{L}{30}u_i^m v_j^m - \frac{L}{15}u_j^m v_i^m - \frac{L}{20}u_i^m v_i^m, & \frac{(b+1)L}{3} - \frac{L}{20}u_j^m v_i^m - \frac{L}{10}u_i^m v_j^m - \frac{L}{5}u_j^m v_j^m \end{bmatrix} * \quad (3.27)$$

$$* \begin{bmatrix} u_i^m \\ u_j^m \end{bmatrix} - \frac{aL}{2} \begin{bmatrix} 1 \\ 1 \end{bmatrix},$$

and the time residuals ($[R_i]_{time}^{(e)}$) and ($[R_j]_{time}^{(e)}$) give :

$$\begin{bmatrix} R_i \\ R_j \end{bmatrix}_{time}^{(e)} = \begin{bmatrix} \frac{L}{3\Delta t} & \frac{L}{6\Delta t} \\ \frac{L}{6\Delta t} & \frac{L}{3\Delta t} \end{bmatrix}^{(e)} \begin{bmatrix} u_i^{m+1} \\ u_j^{m+1} \end{bmatrix} + \begin{bmatrix} -\frac{L}{3\Delta t} & -\frac{L}{6\Delta t} \\ -\frac{L}{6\Delta t} & -\frac{L}{3\Delta t} \end{bmatrix}^{(e)} \begin{bmatrix} u_i^{m} \\ u_j^{m} \end{bmatrix}. \tag{3.28}$$

Each has a set of terms evaluated at time step $(m+1)$ and another at time step(m). Combining the terms at time step (m+1) gives the element matrix:

$$[B]^{(e)} = \theta\left(\frac{d_1}{L}\right)^{(e)} \begin{bmatrix} 1 & -1 \\ -1 & 1 \end{bmatrix} \begin{bmatrix} u_i^{m+1} \\ u_j^{m+1} \end{bmatrix} + \left(\frac{L}{6\Delta t}\right)^{(e)} \begin{bmatrix} 2 & 1 \\ 1 & 2 \end{bmatrix} \begin{bmatrix} u_i^{m+1} \\ u_j^{m+1} \end{bmatrix}. \tag{3.29}$$

Also the element column is :

$$[C]^{(e)} = -(1-\theta)\left(\frac{d_1}{L}\right)^{(e)} \begin{bmatrix} 1 & -1 \\ -1 & 1 \end{bmatrix} \begin{bmatrix} u_i^{m} \\ u_j^{m} \end{bmatrix} + \left(\frac{L}{6\Delta t}\right)^{(e)} \begin{bmatrix} 2 & 1 \\ 1 & 2 \end{bmatrix} \begin{bmatrix} u_i^{m} \\ u_j^{m} \end{bmatrix} \tag{3.30}$$

$$[C]^{(e)} = \begin{bmatrix} \alpha & \beta \\ \gamma & \delta \end{bmatrix} \begin{bmatrix} u_i^{m} \\ u_j^{m} \end{bmatrix} - \frac{aL}{2} \begin{bmatrix} 1 \\ 1 \end{bmatrix},$$

where

$$\alpha = -(1-\theta)\frac{d_1}{L} + \frac{2L}{6\Delta t} - \frac{(b+1)L}{3} + \frac{L}{5}u_i^m v_i^m + \frac{L}{10}u_j^m v_i^m + \frac{L}{20}u_i^m v_j^m,$$

$$\beta = (1-\theta)\frac{d_1}{L} + \frac{L}{6\Delta t} - \frac{(b+1)L}{6} + \frac{L}{30}u_j^m v_i^m + \frac{L}{15}u_i^m v_j^m + \frac{L}{20}u_j^m v_j^m,$$

$$\gamma = (1-\theta)\frac{d_1}{L} + \frac{L}{6\Delta t} - \frac{(b+1)L}{6} + \frac{L}{30}u_i^m v_j^m + \frac{L}{15}u_j^m v_i^m + \frac{L}{20}u_i^m v_i^m,$$

$$\delta = -(1-\theta)\frac{d_1}{L} + \frac{2L}{6\Delta t} - \frac{(b+1)L}{3} + \frac{L}{5}u_j^m v_j^m + \frac{L}{10}u_i^m v_j^m + \frac{L}{20}u_j^m v_i^m.$$

Also we can write the second equation of the system as :

$$\begin{bmatrix} O_i \\ O_j \end{bmatrix}^{(e)} = [B]^{(e)} \begin{bmatrix} v_i^{m+1} \\ v_j^{m+1} \end{bmatrix} - [C]^{(e)} \begin{bmatrix} v_i^{m} \\ v_j^{m} \end{bmatrix},$$

where

$$[B]^{(e)} = \theta\left(\frac{d_2}{L}\right)^{(e)} \begin{bmatrix} 1 & -1 \\ -1 & 1 \end{bmatrix} \begin{bmatrix} v_i^{m+1} \\ v_j^{m+1} \end{bmatrix} + \left(\frac{L}{6\Delta t}\right)^{(e)} \begin{bmatrix} 2 & 1 \\ 1 & 2 \end{bmatrix} \begin{bmatrix} v_i^{m+1} \\ v_j^{m+1} \end{bmatrix} \tag{3.31}$$

Also the element column is:

$$[C]^{(e)}=-(1-\theta)\left(\frac{d_2}{L}\right)^{(e)}\begin{bmatrix}1 & -1\\ -1 & 1\end{bmatrix}\begin{bmatrix}v_i^m\\ v_j^m\end{bmatrix}+\left(\frac{L}{6\Delta t}\right)^{(e)}\begin{bmatrix}2 & 1\\ 1 & 2\end{bmatrix}\begin{bmatrix}v_i^m\\ v_j^m\end{bmatrix}$$

$$-\begin{bmatrix}\frac{L}{5}(u_i^{m+1})^2+\frac{L}{10}u_i^{m+1}u_j^{m+1}+\frac{L}{30}(u_j^m)^2 & \frac{L}{20}(u_i^{m+1})^2+\frac{L}{15}u_i^{m+1}u_j^{m+1}+\frac{L}{20}(u_j^{m+1})\\ \frac{L}{20}(u_i^{m+1})^2+\frac{L}{15}u_i^{m+1}u_j^{m+1}+\frac{L}{20}(u_j^{m+1})^2 & \frac{L}{30}(u_i^{m+1})^2+\frac{L}{10}u_i^{m+1}u_j^{m+1}+\frac{L}{5}(u_j^{m+1})^2\end{bmatrix}\begin{bmatrix}v_i^m\\ v_j^m\end{bmatrix} \quad (3.32)$$

$$-\frac{bL}{6}\begin{bmatrix}2 & 1\\ 1 & 2\end{bmatrix}\begin{bmatrix}u_i^{m+1}\\ u_j^{m+1}\end{bmatrix}.$$

Assembling the global matrices for the equation $[B]^{(e)}$ and $[C]^{(e)}$ for the first equation of (1.1) of the system, we get :

$$\begin{bmatrix}
[\frac{\theta d_1}{L}+\frac{L}{3\Delta t}]^{(e)} & [-\frac{\theta d_1}{L}+\frac{L}{6\Delta t}]^{(e)} & 0 & 0 & & 0 \ldots\ldots 0\\
[-\frac{\theta d_1}{L}+\frac{L}{6\Delta t}]^{(e)} & [\frac{2\theta d_1}{L}+\frac{2L}{3\Delta t}]^{(e)} & [-\frac{\theta d_1}{L}+\frac{L}{6\Delta t}]^{(e)} & 0 & & 0\ldots\ldots 0\\
0 & [-\frac{\theta d_1}{L}+\frac{L}{6\Delta t}]^{(e)} & [\frac{2\theta d_1}{L}+\frac{2L}{3\Delta t}]^{(e)} & [-\frac{\theta d_1}{L}+\frac{L}{6\Delta t}]^{(e)} & & 0\ldots 0\\
. & . & . & . & & \\
. & 0\ldots & 0 & 0 & \ldots\; 0\ldots\; 0 & \\
0\;\ldots\ldots & 0\ldots\ldots\ldots & 0\;\ldots\ldots\;[-\frac{\theta d_1}{L}+\frac{L}{6\Delta t}]^{(e)} & [\frac{2\theta d_1}{L}+\frac{2L}{3\Delta t}]^{(e)} & [-\frac{\theta d_1}{L}+\frac{L}{6\Delta t}]^{(e)} & \\
0\ldots\ldots\ldots & 0\ldots\ldots\ldots\ldots 0\ldots\ldots & & [-\frac{\theta d_1}{L}+\frac{L}{6\Delta t}]^{(e)} & \frac{\theta d_1}{L}+\frac{L}{3\Delta t}]^{(e)} &
\end{bmatrix} *$$

$$\begin{bmatrix}u_1^{m+1}\\ u_2^{m+1}\\ u_3^{m+1}\\ \vdots\\ \vdots\\ \vdots\\ u_{n-1}^{m+1}\\ u_n^{m+1}\end{bmatrix}=\begin{bmatrix}a_{11}\\ a_{21}\\ a_{31}\\ \vdots\\ \vdots\\ \vdots\\ a_{(n-1)1}\\ a_{n1}\end{bmatrix}, \quad (3.33)$$

Where

$$a_{11} = [\left(-(1-\theta)\frac{d_1}{L} + \frac{2L}{6\Delta t} - (b+1)\frac{L}{3} + \frac{L}{5}u_1^m v_1^m + \frac{L}{10}u_2^m v_1^m + \frac{L}{20}u_1^m v_2^m \right)u_1 +$$

$$+\left((1-\theta)\frac{d_1}{L} + \frac{L}{6\Delta t} - (b+1)\frac{L}{6} + \frac{L}{30}u_2^m v_1^m + \frac{L}{15}u_1^m v_2^m + \frac{L}{20}u_2^m v_2^m \right)u_2] - \frac{aL}{2},$$

$$a_{21} = [\left((1-\theta)\frac{d_1}{L} + \frac{L}{6\Delta t} - (b+1)\frac{L}{6} + \frac{L}{30}u_1^m v_2^m + \frac{L}{15}u_2^m v_1^m + \frac{L}{20}u_1^m v_1^m \right)u_1^m +$$

$$+\left(-(1-\theta)\frac{d_1}{L} + \frac{2L}{6\Delta t} - (b+1)\frac{L}{3} + + \frac{L}{20}u_2^m v_1^m + \frac{L}{10}u_1^m v_2^m + \frac{L}{5}u_2^m v_2^m \right)u_2^m] - \frac{aL}{2} +$$

$$+[\left(-(1-\theta)\frac{d_1}{L} + \frac{2L}{6\Delta t} - (b+1)\frac{L}{3} + \frac{L}{5}u_2^m v_2^m + \frac{L}{10}u_3^m v_2^m + \frac{L}{20}u_2^m v_3^m \right)u_2^m +$$

$$+\left((1-\theta)\frac{d_1}{L} + \frac{L}{6\Delta t} - (b+1)\frac{L}{6} + \frac{L}{30}u_3^m v_2^m + \frac{L}{15}u_2^m v_3^m + \frac{L}{20}u_3^m v_3^m \right)u_3^m] - \frac{aL}{2},$$

$$a_{31} = [\left((1-\theta)\frac{d_1}{L} + \frac{L}{6\Delta t} - (b+1)\frac{L}{6} + \frac{L}{30}u_2^m v_3^m + \frac{L}{15}u_3^m v_2^m + \frac{L}{20}u_2^m v_2^m \right)u_2^m +$$

$$+\left(-(1-\theta)\frac{d_1}{L} + \frac{2L}{6\Delta t} - (b+1)\frac{L}{3} + \frac{L}{20}u_3^m v_2^m + \frac{L}{10}u_2^m v_3^m + \frac{L}{5}u_3^m v_3^m \right)u_3^m] - \frac{aL}{2} +$$

$$+[\left(-(1-\theta)\frac{d_1}{L} + \frac{2L}{6\Delta t} - (b+1)\frac{L}{3} + \frac{L}{5}u_3^m v_3^m + \frac{L}{10}u_4^m v_3^m + \frac{L}{20}u_3^m v_4^m \right)u_3^m +$$

$$+\left((1-\theta)\frac{d_1}{L} + \frac{L}{6\Delta t} - (b+1)\frac{L}{6} + \frac{L}{30}u_4^m v_3^m + \frac{L}{15}u_3^m v_4^m + \frac{L}{20}u_4^m v_4^m \right)u_4^m] - \frac{aL}{2},$$

$$\vdots \qquad\qquad .$$

$$a_{(n-1)1} = [\left((1-\theta)\frac{d_1}{L} + \frac{L}{6\Delta t} - (b+1)\frac{L}{6} + \frac{L}{30}u_{n-2}^m v_{n-1}^m + \frac{L}{15}u_{n-1}^m v_{n-2}^m + \frac{L}{20}u_{n-2}^m v_{n-2}^m \right)u_{n-2}^m +$$

$$+\left(-(1-\theta)\frac{d_1}{L} + \frac{2L}{6\Delta t} - (b+1)\frac{L}{3} + \frac{L}{20}u_{n-1}^m v_{n-2}^m + \frac{L}{10}u_{n-2}^m v_{n-1}^m + \frac{L}{5}u_{n-1}^m v_{n-1}^m \right)u_{n-1}^m] - aL/2 +$$

$$+[\left(-(1-\theta)\frac{d_1}{L} + \frac{2L}{6\Delta t} - (b+1)\frac{L}{3} + \frac{L}{5}u_{n-1}^m v_{n-1}^m + \frac{L}{10}u_n^m v_{n-1}^m + \frac{L}{20}u_{n-1}^m v_n^m \right)u_{n-1}^m +$$

$$+\left((1-\theta)\frac{d_1}{L} + \frac{L}{6\Delta t} - (b+1)\frac{L}{6} + \frac{L}{30}u_n^m v_{n-1}^m + \frac{L}{15}u_{n-1}^m v_n^m + \frac{L}{20}u_n^m v_n^m \right)u_n^m] - \frac{aL}{2},$$

$$a_{n1} = [\left((1-\theta)\frac{d_1}{L} + \frac{L}{6\Delta t} - (b+1)\frac{L}{6} + \frac{L}{30}u_{n-1}^m v_n^m + \frac{L}{15}u_n^m v_{n-1}^m + \frac{L}{20}u_{n-1}^m v_{n-1}^m \right)u_{n-1}^m +$$

$$+\left((-(1-\theta)\frac{d_1}{L} + \frac{2L}{6\Delta t} - (b+1)\frac{L}{3} + \frac{L}{20}u_n^m v_{n-1}^m + \frac{L}{10}u_{n-1}^m v_n^m + \frac{L}{5}u_n^m v_n^m) \right)u_n^n] - \frac{aL}{2},$$

We can solve the previous system by using the Gaussian elimination method.

Assembling the global matrices for the equations ($[B]^{(e)}$) and ($[C]^{(e)}$) for the second equation of the system we get :

$$\begin{bmatrix} [\frac{\theta d_2}{L}+\frac{L}{3\Delta t}]^{(e)} & [-\frac{\theta d_2}{L}+\frac{L}{6\Delta t}]^{(e)} & 0 & 0 & & 0 \ldots\ldots 0 \\ [-\frac{\theta d_2}{L}+\frac{L}{6\Delta t}]^{(e)} & [\frac{2\theta d_2}{L}+\frac{2L}{3\Delta t}]^{(e)} & [-\frac{\theta d_2}{L}+\frac{L}{6\Delta t}]^{(e)} & 0 & & 0\ldots\ldots 0 \\ 0 & [-\frac{\theta d_2}{L}+\frac{L}{6\Delta t}]^{(e)} & [\frac{2\theta d_2}{L}+\frac{2L}{3\Delta t}]^{(e)} & [-\frac{\theta d_2}{L}+\frac{L}{6\Delta t}]^{(e)} & & 0\ldots 0 \\ . & . & . & . & & \\ . & 0\ldots & 0 & 0 & \ldots\ 0\ldots & 0 \\ 0\ \ldots\ldots & 0\ldots\ldots\ldots\ldots & 0\ \ldots\ldots\ [-\frac{\theta d_2}{L}+\frac{L}{6\Delta t}]^{(e)} & [\frac{2\theta d_2}{L}+\frac{2L}{3\Delta t}]^{(e)} & [-\frac{\theta d_2}{L}+\frac{L}{6\Delta t}]^{(e)} & \\ 0\ldots\ldots\ldots\ldots & 0\ldots\ldots\ldots\ldots\ldots\ldots 0\ldots\ldots & & [-\frac{\theta d_2}{L}+\frac{L}{6\Delta t}]^{(e)} & \frac{\theta d_2}{L}+\frac{L}{3\Delta t}]^{(e)} & \end{bmatrix} *$$

$$\begin{bmatrix} v_1^{m+1} \\ v_2^{m+1} \\ v_3^{m+1} \\ \vdots \\ \vdots \\ v_{n-1}^{m+1} \\ v_n^{m+1} \end{bmatrix} = \begin{bmatrix} b_{11} \\ b_{21} \\ b_{31} \\ \vdots \\ \vdots \\ b_{(n-1)1} \\ b_{n1} \end{bmatrix}, \qquad (3.34)$$

Where

$$b_{11} = (-(1-\theta)\frac{d_2}{L} + \frac{2L}{6\Delta t} - \frac{L}{5}(u_1^{m+1})^2 - \frac{L}{10}u_1^{m+1}u_2^{m+1} - \frac{L}{30}(u_2^{m+1})^2)v_1^m - \frac{2bL}{6}u_1^{m+1}$$
$$+((1-\theta)\frac{d_2}{L} + \frac{L}{6\Delta t} - \frac{L}{20}(u_1^{m+1})^2 - \frac{L}{15}u_1^{m+1}u_2^{m+1} - \frac{L}{20}(u_2^{m+1})^2)v_2^m - \frac{bL}{6}u_2^{m+1},$$

$$b_{21} = ((1-\theta)\frac{d_2}{L} + \frac{L}{6\Delta t} - \frac{L}{20}(u_1^{m+1})^2 - \frac{L}{15}u_1^{m+1}u_2^{m+1} - \frac{L}{20}(u_2^{m+1})^2)v_1^m - \frac{bL}{6}u_1^{m+1}$$

$$+(-(1-\theta)\frac{d_2}{L} + \frac{2L}{6\Delta t} - \frac{L}{30}(u_1^{m+1})^2 - \frac{L}{10}u_1^{m+1}u_2^{m+1} - \frac{L}{5}(u_2^{m+1})^2)v_2^m - \frac{2bL}{6}u_2^{m+1} +$$

$$+(-(1-\theta)\frac{d_2}{L} + \frac{2L}{6\Delta t} - \frac{L}{5}(u_2^{m+1})^2 - \frac{L}{30}(u_3^{m+1})^2 - \frac{L}{10}u_2^{m+1}u_3^{m+1})v_2^m - \frac{2bL}{6}u_2^{m+1} +$$

$$((1-\theta)\frac{d_2}{L} + \frac{L}{6\Delta t} - \frac{L}{20}(u_2^{m+1})^2 - \frac{L}{15}u_2^{m+1}u_3^{m+1} - \frac{L}{20}(u_3^{m+1})^2)v_3^m - \frac{bL}{6}u_2^{m+1} ,$$

$$b_{31} = ((1-\theta)\frac{d_2}{L} + \frac{L}{6\Delta t} - \frac{L}{20}(u_2^{m+1})^2 - \frac{L}{15}u_2^{m+1}u_3^{m+1} - \frac{L}{20}(u_3^{m+1})^2)v_2^m - \frac{bL}{6}u_2^{m+1} +$$

$$+(-(1-\theta)\frac{d_2}{L} + \frac{2L}{6\Delta t} - \frac{L}{30}(u_2^{m+1})^2 - \frac{L}{10}u_2^{m+1}u_3^{m+1} - \frac{L}{5}(u_3^{m+1})^2)v_3^m - \frac{2bL}{6}u_3^{m+1} +$$

$$+(-(1-\theta)\frac{d_2}{L} + \frac{2L}{6\Delta t} - \frac{L}{5}(u_3^{m+1})^2 - \frac{L}{10}u_3^{m+1}u_4^{m+1} - \frac{L}{30}(u_4^{m+1})^2)v_3^m - \frac{2bL}{6}u_3^{m+1} +$$

$$+((1-\theta)\frac{d_2}{L} + \frac{L}{6\Delta t} - \frac{L}{20}(u_3^{m+1})^2 - \frac{L}{15}u_3^{m+1}u_4^{m+1} - \frac{L}{20}(u_4^{m+1})^2)v_4^m - \frac{bL}{6}u_4^{m+1} ,$$

$$\vdots$$

$$b_{(n-1)1} = ((1-\theta)\frac{d_2}{L} + \frac{L}{6\Delta t} - \frac{L}{20}(u_{n-2}^{m+1})^2 - \frac{L}{15}u_{n-1}^{m+1}u_{n-2}^{m+1} - \frac{L}{20}(u_{n-1}^{m+1})^2)v_{n-2}^m - \frac{bL}{6}u_{n-2}^{m+1} +$$

$$+(-(1-\theta)\frac{d_2}{L} + \frac{2L}{6\Delta t} - \frac{L}{30}(u_{n-2}^{m+1})^2 - \frac{L}{10}u_{n-2}^{m+1}u_{n-1}^{m+1} - \frac{L}{5}(u_{n-1}^{m+1})^2)v_{n-1}^m - \frac{2bL}{6}u_{n-1}^{m+1} +$$

$$+(-(1-\theta)\frac{d_2}{L} + \frac{2L}{6\Delta t} - \frac{L}{5}(u_{n-1}^{m+1})^2 - \frac{L}{10}u_{n-1}^{m+1}u_n^{m+1} - \frac{L}{30}(u_n^{m+1})^2)v_{n-1}^m - \frac{2bL}{6}u_{n-1}^{m+1} +$$

$$+((1-\theta)\frac{d_2}{L} + \frac{L}{6\Delta t} - \frac{L}{20}(u_{n-1}^{m+1})^2 - \frac{L}{15}u_{n-1}^{m+1}u_n^{m+1} - \frac{L}{20}(u_n^{m+1})^2 +)v_n^m - \frac{bL}{6}u_n^{m+1} ,$$

$$b_{n1} = ((1-\theta)\frac{d_2}{L} + \frac{L}{6\Delta t} - \frac{L}{20}(u_{n-1}^{m+1})^2 - \frac{L}{15}u_{n-1}^{m+1}u_n^{m+1} - \frac{L}{20}(u_n^{m+1})^2)v_{n-1}^m - \frac{bL}{6}u_{n-1}^{m+1}$$

$$+(-(1-\theta)\frac{d_2}{L} + \frac{2L}{6\Delta t} - \frac{L}{30}(u_{n-1}^{m+1})^2 - \frac{L}{10}u_{n-1}^{m+1}u_n^{m+1} - \frac{L}{5}(u_n^{m+1})^2)v_n^m - \frac{2bL}{6}u_n^{m+1}.$$

We can solve the systems (3.33) and (3.34) by using the Gaussian elimination method.

CHAPTER FOUR

Expansion Methods

4.1 Introduction

One of the uses of approximating functions is to replace complicated functions by some simpler functions so that many operations such as integration can be easily performed. In this chapter, a solution to the Brusselator system (1.1) is approximated by polynomials in two independent variables. In two dimensions, a complete n^{th} degree polynomial is given by :

$$u_i(x,t) \approx P_{in}(x,t) = \sum_{\substack{j=0 \\ j+k \le n}}^{n} \sum_{k=0}^{n} c_{ijk}\, x^j t^k, \quad i=1, 2,...,N \qquad (4.1)$$

where j and k are permuted accordingly. The number of terms in the above polynomials is equal to $\frac{(n+1)\times(n+2)}{2}$.

For example:

if $n=1$, $\quad P_{i1}(x,t) = c_{i00} + c_{i01}t + c_{i10}x,$ (3 terms)

if $n=2$, $\quad P_{i2}(x,t) = c_{i00} + c_{i01}t + c_{i02}t^2 + c_{i10}x + c_{i11}xt + c_{i20}x^2,$ (6 terms)

if n=3, $\quad P_{i3}(x,t) = c_{i00} + c_{i01}t + c_{i02}t^2 + c_{i03}t^3 + c_{i10}x + c_{i11}xt + c_{i12}xt^2 + c_{i20}x^2 + c_{i21}x^2t + c_{i30}x^3$ (10 terms)

In order to consolidate the expansion method, some error minimizing technique to determine the coefficients $\underset{j+k\le n}{c_{ijk}}$, $i=1, 2,\ldots,N$; j, $k=0, 1,...,n$ in (4.1) are needed. One of the most popular minimizing techniques is the weighted residual method (WRM) [32] , [63] which includes the {collocation method (CM), partition method (PM), Galerkin method (GM) and Least-Square Method (LSM)}. In this chapter, we choose the weighted residual methods for finding the unknown coefficient.

4.2 Formulation of WRM's for Solving Brusselator System

The WRM's are formulated to be suitable for solving Brusselator system as follows:

Consider the functional equation given by

$$T(u_i(x,t)) = p(x) \quad (x,t) \in D, \; i=1, 2, \ldots, N. \tag{4.2}$$

Now by substituting the approximated solution of $u_i(x,t)$ given by (4.1) into the equation (4.2), the result is so called "residual function" defined by :

$$R_{in}(\underset{j+k\le n}{c_{ijk}}, x,t) = T(u_i(x,t)) - p(x), \tag{4.3}$$

where $i=1, 2, \ldots, N$; $j, k=0, 1,\ldots, n$.

The residue $R_{in}(\underset{j+k\le n}{c_{ijk}}, x,t)$ depends on x, t as well as the way that the parameters $\underset{j+k\le n}{c_{ijk}}$, $i=1, 2, \ldots, N$; $j, k=0, 1,\ldots, n$ are chosen. Since the residual function $R_{in}(\underset{j+k\le n}{c_{ijk}}, x,t)$ is identically equal to zero for the exact solution, the goal is to choose the coefficients $\underset{j+k\le n}{c_{ijk}}$ so that the residual function can be minimized.

The main aim for using the *WRM*'s is to find the coefficients $\underset{j+k\le n}{c_{ijk}}$ so that the residue $R_{in}(\underset{j+k\le n}{c_{ijk}}, x,t)$ becomes small (in fact zero) over a chosen domain. In integral form, this can be achieved with the condition

$$\iint_D w_{jk}(x,t) R_{in}(\underset{j+k\le n}{c_{ijk}}, x, t)\,dt\,dx = 0, \tag{4.4}$$

where $w_{jk}(x,t)$ are prescribed weight function. The technique described by (4.4) is called *WRM*'s by which the optimal values of $L=\dfrac{N\times(n+1)\times(n+2)}{2}$ coefficients $\underset{j+k\le n}{c_{ijk}}$ that minimize $R_{in}(\underset{j+k\le n}{c_{ijk}}, x,t)$, are obtained.

We now present four methods of the *WRM*'s to determine $\underset{j+k\le n}{c_{ijk}}$ in the equation (4.1) as follows:

4.2.1 Collocation Method

The main idea behind this method is the parameters $\underset{j+k\le n}{c_{ijk}}$, $i=1, 2,\ldots,N$; $j, k=0, 1,\ldots,n$ that are to be found by requiring that the residual $R_{in}(\underset{j+k\le n}{c_{ijk}}, x,t)$ be zero at the given sets of the collocation points $x_0, x_1, ..., x_n$ and $t_0, t_1, ..., t_n$ on the interval $[a_1,b_1]$ and $[a_2,b_2]$ respectively, where $x_j = a_1 + jh_x$, $t_k = a_2 + kh_t$; j, k=0, 1,...,n, $h_x = \frac{b_1 - a_1}{n}$ and $h_t = \frac{b_2 - a_2}{n}$ as follows:

In this method the weighted functions will be Dirac delta functions in two dimensional Cartesian coordinates [25], i.e.

$$w_{jk}(x,t) = \delta^2(x,t) = \delta(x - x_j)\delta(t - t_k),$$

which vanishes everywhere except at $x = x_j$, $j = 0,1,...,n$ and $t = t_k$, $k = 0,1,...,n$ such that $j + k \le n$. This means that

$$\delta(x - x_j) = \begin{cases} 0 & \text{if } x \ne x_j \\ 1 & \text{if } x = x_j \end{cases} \quad \text{for } j = 0, 1, ..., n$$

and

$$\delta(t - t_k) = \begin{cases} 0 & \text{if } t \ne t_k \\ 1 & \text{if } t = t_k \end{cases} \quad \text{for } k = 0, 1, ..., n.$$

Then, the equation (4.4) becomes :

$$\iint_D w_{jk}(x,t) R_{in}(\underset{j+k\le n}{c_{ijk}}, x, t) dt dx = \iint_D \delta(x - x_j)\delta(t - t_k) R_{in}[\underset{j+k\le n}{c_{ijk}}, x, t] dt dx = 0,$$

this can be written as :

$$R_{in}(\underset{j+k\le n}{c_{ijk}}, x_r, t_q) = 0, i = 1, 2, ..., N;\ r, q = 0, 1, ..., n \text{ such that } r+q \le n. \quad (4.5)$$

From equation (4.5), we get an $L \times L$ nonlinear system of simultaneous equations for the coefficients $\underset{j+k\le n}{c_{ijk}}$, where $L = \frac{N \times (n+1) \times (n+2)}{2}$, which can be solved by modified Newton-Raphson method for finding the parameters $\underset{j+k\le n}{c_{ijk}}$.

4.2.2 Sub-Domain Method

The idea is to force the weighted residual equal to zero not just at fixed points in the domain, but over various subsections of the domain. To accomplish this, the weight functions are set to unity, and the integral over the entire domain is broken into a number of sub-domains sufficient to evaluate all unknown parameters $\underset{j+k\le n}{c_{ijk}}$ as follows:

First, divide each of the intervals $[a_1, b_1]$ and $[a_2, b_2]$ into $n+1$ sub-intervals $D_j=[x_j,x_{j+1}]$ and $D_k=[t_k,t_{k+1}]$ respectively where $x_j=a_1+jh_x$, $t_k=a_2+kh_t$; j, $k=0, 1,\ldots,n$, $h_x=\dfrac{b_1-a_1}{n+1}$, $h_t=\dfrac{b_2-a_2}{n+1}$, and

$$w_{jk}(x,t)=\begin{cases}1 & if\ \ x\in D_j\ \ and\ \ t\in D_k\\ 0 & if\ \ x\notin D_j\ \ or\ t\notin D_k\end{cases};\quad j, k=0, 1,\ldots, n$$

Second, integrate the residual (4.3) with respect to x and t such that for each sub-intervals D_j we must use all sub-intervals D_k provided that the sum of indexes of the lower limits of the integrals are less than or equal to N. Hence, the equation (4.4) becomes :

$$\int_{x_j}^{x_{j+1}}\int_{t_k}^{t_{k+1}} R_{in}(\underset{j+k\le n}{c_{ijk}}, x,t)dtdx=0,\quad i=1, 2,\ldots N;\ j, k=0, 1,\ldots,n \text{ such that } j+k\le n$$

This means that:

$$\left.\begin{array}{l}
\displaystyle\int_{x_0}^{x_1}\int_{t_v}^{t_{v+1}} R_{in}(\underset{j+k\le n}{c_{ijk}}, x,t)dtdx=0, \text{ for } i=1, 2, \ldots,N \text{ and } v=0, 1,\ldots,n\\
\displaystyle\int_{x_1}^{x_2}\int_{t_v}^{t_{v+1}} R_{in}(\underset{j+k\le n}{c_{ijk}}, x,t)dtdx=0, \text{ for } i=1, 2, \ldots, N \text{ and } v=0, 1,\ldots,n-1\\
\displaystyle\int_{x_2}^{x_3}\int_{t_v}^{t_{v+1}} R_{in}(\underset{j+k\le n}{c_{ijk}}, x,t)dtdx=0, \text{ for } i=1, 2, \ldots, N \text{ and } v=0, 1,\ldots,n-2\\
\vdots\\
\displaystyle\int_{x_N}^{x_{N+1}}\int_{t_v}^{t_{v+1}} R_{in}(\underset{j+k\le n}{c_{ijk}}, x,t)dtdx=0, \text{ for } i=1, 2, \ldots, N \text{ and } v=0.
\end{array}\right\}\quad (4.6)$$

Equation (4.6) yields an $L\times L$ nonlinear system of simultaneous equations for the coefficients $\underset{j+k\le n}{c_{ijk}}$. Solve this system by Modified Newton-Raphson method for finding the parameters $\underset{j+k\le n}{c_{ijk}}$.

4.2.3 Least Square Method

The continuous summation of all the squared residual is minimized; the rationale behind the name can be seen. In other words, a minimum of

$$S=\iint_D R_{in}^2(\underset{j+k\le n}{c_{ijk}}, x, t)\ dtdx .$$

in order to achieve a minimum of this scalar function, the derivatives of S with respect to all the unknown parameters must be zero. That is,

$$\frac{\partial S}{\underset{r+q\le n}{\partial c_{irq}}} = 0 = 2\iint_D R_{in}(\underset{j+k\le n}{c_{ijk}}, x, t)\frac{\partial R_{in}(\underset{j+k\le n}{c_{ijk}}, x, t)}{\underset{r+q\le n}{\partial c_{irq}}}\ dtdx . \qquad (4.7)$$

for i=1, 2, …, N; r,q=0, 1, …, n , such that $r+q\le n$.

Comparing with (4.4), the weight functions are to be :

$$w_{irq}(x,t) = 2\frac{\partial R_{in}(\underset{j+k\le n}{c_{ijk}}, x, t)}{\underset{r+q\le n}{\partial c_{irq}}} .$$

Hence, from the equation (4.7), we get :

$$\int_{a_1}^{b_1}\int_{a_2}^{b_2} R_{in}(\underset{j+k\le n}{c_{ijk}}, x, t)\frac{\partial R_{in}(\underset{j+k\le n}{c_{ijk}}, x, t)}{\underset{r+q\le n}{\partial c_{irq}}} dtdx = 0, \qquad (4.8)$$

for i=1, 2, …, N; r,q=0, 1, …, n such that $r+q\le$n.

The equation (4.8) is also a system of L nonlinear equations in the L unknown coefficients $\underset{j+k\le n}{c_{ijk}}$. Solve this system by modified Newton-Raphson method for finding the parameters $\underset{j+k\le n}{c_{ijk}}$.

4.2.4 Galerkin Method

This method may be viewed as a modification of the LSM rather than using the derivative of the residual with respect to the unknown $\underset{j+k\le n}{c_{ijk}}$, the derivative of the approximating function (4.1) is used [70]. That is, if the function is approximated as in (4.1), then the weight functions are

$$w_{irq}(x,t)=\frac{\partial P_{in}(x,t)}{\partial \underset{r+q\le n}{c_{irq}}},\quad i\text{=}1, 2,\ldots,N \text{ and } r, q\text{=}0, 1,\ldots,n \text{ such that } r+q\le n.$$

Hence, from the equation (5.8), we get :

$$\int_{a_1}^{b_1}\int_{a_2}^{b_2} R_{in}(\underset{j+k\le n}{c_{ijk}}, x, t)\frac{\partial P_{in}(x,t)}{\partial \underset{r+q\le n}{c_{irq}}}\,dtdx=0, \tag{4.9}$$

for i=1, 2, …, N; and r, q=0, 1,…, n such that $r+q\le n$

The above system is also $L\times L$ nonlinear system and we solve it by modified Newton-Raphson method for finding the parameters $\underset{j+k\le n}{c_{ijk}}$.

4.3 Solution of the Brusselator system by Using WRM's

In this section, we approximate the solution of Brusselator system (1.1) by means of the *WRM*'s, described in section 4.2, and we attempt to calculate the coefficients $\underset{j+k\le n}{c_{ijk}}$, i=1, 2,…,N; j, k=0, 1,…,n as follows:

Using (4.1) and (4.3), we can write the residual equation (4.3) for (1.1) in the form

$$\left.\begin{aligned}
&R_{1n}(\underset{j+k\le n}{c_{1jk}}, x, t)=\frac{\partial}{\partial t}\left(\sum_{\substack{j=0\\ j+k\le n}}^{n}\sum_{k=0}^{n} c_{1jk}\,x^j t^k\right)-d_1\frac{\partial^2}{\partial x^2}\left(\sum_{\substack{j=0\\ j+k\le n}}^{n}\sum_{k=0}^{n} c_{1jk}\,x^j t^k\right)-\\
&\left(\sum_{\substack{j=0\\ j+k\le n}}^{n}\sum_{k=0}^{n} c_{1jk}\,x^j t^k\right)^2\left(\sum_{\substack{j=0\\ j+k\le n}}^{n}\sum_{k=0}^{n} c_{2jk}\,x^j t^k\right)+(b+1)\sum_{\substack{j=0\\ j+k\le n}}^{n}\sum_{k=0}^{n} c_{1jk}\,x^j t^k-a,\ R_{2n}(\underset{j+k\le n}{c_{2jk}}, x, t)=\\
&\frac{\partial}{\partial t}\left(\sum_{\substack{j=0\\ j+k\le n}}^{n}\sum_{k=0}^{n} c_{2jk}\,x^j t^k\right)-d_2\frac{\partial^2}{\partial x^2}\left(\sum_{\substack{j=0\\ j+k\le n}}^{n}\sum_{k=0}^{n} c_{2jk}\,x^j t^k\right)\\
&+\left(\sum_{\substack{j=0\\ j+k\le n}}^{n}\sum_{k=0}^{n} c_{1jk}\,x^j t^k\right)^2\left(\sum_{\substack{j=0\\ j+k\le n}}^{n}\sum_{k=0}^{n} c_{2jk}\,x^j t^k\right)-b\sum_{\substack{j=0\\ j+k\le n}}^{n}\sum_{k=0}^{n} c_{1jk}\,x^j t^k.
\end{aligned}\right\}\tag{4.10}$$

Hereunder, we try to find the parameters $\underset{j+k\le n}{c_{ijk}}$, i=1,2 using the methods described in section 4.2 as follows:

4.3.1 Collocation Method

From the equations (4.10) and (4.5) we get

$$
\begin{aligned}
&\frac{\partial}{\partial t}\left(\sum_{\substack{j=0 \\ j+k\le n}}^{n}\sum_{k=0}^{n} c_{1jk}\, x_o^j t^k\right)\Bigg|_{t=t_q} - d_1 \frac{\partial^2}{\partial x^2}\left(\sum_{\substack{j=0 \\ j+k\le n}}^{n}\sum_{k=0}^{n} c_{1jk}\, x^j t_q^k\right)\Bigg|_{x=x_0} - \\
&\left(\sum_{\substack{j=0 \\ j+k\le n}}^{n}\sum_{k=0}^{n} c_{1jk}\, x_0^j t_q^k\right)^2\left(\sum_{\substack{j=0 \\ j+k\le n}}^{n}\sum_{k=0}^{n} c_{2jk}\, x_0^j t_q^k\right) + (b+1)\sum_{\substack{j=0 \\ j+k\le n}}^{n}\sum_{k=0}^{n} c_{1jk}\, x_0^j t_q^k = a, \\
&\frac{\partial}{\partial t}\left(\sum_{\substack{j=0 \\ j+k\le n}}^{n}\sum_{k=0}^{n} c_{2jk}\, x_0^j t^k\right)\Bigg|_{t=t_q} - d_2 \frac{\partial^2}{\partial x^2}\left(\sum_{\substack{j=0 \\ j+k\le n}}^{n}\sum_{k=0}^{n} c_{2jk}\, x^j t_q^k\right)\Bigg|_{x=x_0} \\
&+\left(\sum_{\substack{j=0 \\ j+k\le n}}^{n}\sum_{k=0}^{n} c_{1jk}\, x_0^j t_q^k\right)^2\left(\sum_{\substack{j=0 \\ j+k\le n}}^{n}\sum_{k=0}^{n} c_{2jk}\, x_0^j t_q^k\right) - b\sum_{\substack{j=0 \\ j+k\le n}}^{n}\sum_{k=0}^{n} c_{1jk}\, x_0^j t_q^k = 0, \\
&\qquad \text{for } q=0, 1, \ldots, n \\
&\frac{\partial}{\partial t}\left(\sum_{\substack{j=0 \\ j+k\le n}}^{n}\sum_{k=0}^{n} c_{1jk}\, x_1^j t^k\right)\Bigg|_{t=t_q} - d_1 \frac{\partial^2}{\partial x^2}\left(\sum_{\substack{j=0 \\ j+k\le n}}^{n}\sum_{k=0}^{n} c_{1jk}\, x^j t_q^k\right)\Bigg|_{x=x_1} - \\
&\left(\sum_{\substack{j=0 \\ j+k\le n}}^{n}\sum_{k=0}^{n} c_{1jk}\, x_1^j t_q^k\right)^2\left(\sum_{\substack{j=0 \\ j+k\le n}}^{n}\sum_{k=0}^{n} c_{2jk}\, x_1^j t_q^k\right) + (b+1)\sum_{\substack{j=0 \\ j+k\le n}}^{N}\sum_{k=0}^{N} c_{1jk}\, x_1^j t_q^k = a, \\
&\frac{\partial}{\partial t}\left(\sum_{\substack{j=0 \\ j+k\le n}}^{n}\sum_{k=0}^{n} c_{2jk}\, x_1^j t^k\right)\Bigg|_{t=t_q} - d_2 \frac{\partial^2}{\partial x^2}\left(\sum_{\substack{j=0 \\ j+k\le n}}^{n}\sum_{k=0}^{n} c_{2jk}\, x^j t_q^k\right)\Bigg|_{x=x_1} \\
&+\left(\sum_{\substack{j=0 \\ j+k\le n}}^{n}\sum_{k=0}^{n} c_{1jk}\, x_1^j t_q^k\right)^2\left(\sum_{\substack{j=0 \\ j+k\le n}}^{n}\sum_{k=0}^{n} c_{2jk}\, x_1^j t_q^k\right) - b\sum_{\substack{j=0 \\ j+k\le n}}^{n}\sum_{k=0}^{n} c_{1jk}\, x_1^j t_q^k = 0,
\end{aligned}
\tag{4.11}
$$

for q=0, 1, …, n-1

⋮

$$\frac{\partial}{\partial t}\left(\sum_{\substack{j=0\\ j+k\le n}}^{n}\sum_{k=0}^{n} c_{1jk}\, x_n^j t^k\right)\Bigg|_{t=t_q} - d_1 \frac{\partial^2}{\partial x^2}\left(\sum_{\substack{j=0\\ j+k\le n}}^{n}\sum_{k=0}^{n} c_{1jk}\, x^j t_q^k\right)\Bigg|_{x=x_n} -$$

$$\left(\sum_{\substack{j=0\\ j+k\le n}}^{n}\sum_{k=0}^{n} c_{1jk}\, x_n^j t_q^k\right)^2\left(\sum_{\substack{j=0\\ j+k\le n}}^{n}\sum_{k=0}^{n} c_{2jk}\, x_n^j t_q^k\right) + (b+1)\sum_{\substack{j=0\\ j+k\le n}}^{n}\sum_{k=0}^{n} c_{1jk}\, x_n^j t_q^k = a,$$

$$\frac{\partial}{\partial t}\left(\sum_{\substack{j=0\\ j+k\le n}}^{n}\sum_{k=0}^{n} c_{2jk}\, x_n^j t^k\right)\Bigg|_{t=t_q} - d_2 \frac{\partial^2}{\partial x^2}\left(\sum_{\substack{j=0\\ j+k\le n}}^{n}\sum_{k=0}^{n} c_{2jk}\, x^j t_q^k\right)\Bigg|_{x=x_n}$$

$$+\left(\sum_{\substack{j=0\\ j+k\le n}}^{n}\sum_{k=0}^{n} c_{1jk}\, x_n^j t_q^k\right)^2\left(\sum_{\substack{j=0\\ j+k\le n}}^{n}\sum_{k=0}^{n} c_{2jk}\, x_n^j t_q^k\right) - b\sum_{\substack{j=0\\ j+k\le n}}^{n}\sum_{k=0}^{n} c_{1jk}\, x_n^j t_q^k = 0,$$

for q=0.

4.3.2 Sub-Domain Method

From the equation (4.10) and (4.6), we get :

$$\int_{x_0}^{x_1}\int_{t_0}^{t_{q+1}}\left\{\frac{\partial}{\partial t}\left(\sum_{\substack{j=0\\ j+k\le n}}^{n}\sum_{k=0}^{n} c_{1jk}\, x^j t^k\right) - d_1\frac{\partial^2}{\partial x^2}\left(\sum_{\substack{j=0\\ j+k\le n}}^{n}\sum_{k=0}^{n} c_{1jk}\, x^j t^k\right) - \left(\sum_{\substack{j=0\\ j+k\le n}}^{n}\sum_{k=0}^{n} c_{1jk}\, x^j t^k\right)^2\left(\sum_{\substack{j=0\\ j+k\le n}}^{n}\sum_{k=0}^{n} c_{2jk}\, x^j t^k\right) + \right.$$

$$\left.(b+1)\sum_{\substack{j=0\\ j+k\le n}}^{n}\sum_{k=0}^{n} c_{1jk}\, x^j t^k\right\} dtdx \qquad = a\int_{x_0}^{x_1}\int_{t_0}^{t_{q+1}} dtdx,$$

$$\int_{x_0}^{x_1}\int_{t_0}^{t_{q+1}}\left\{\frac{\partial}{\partial t}\left(\sum_{\substack{j=0\\ j+k\le n}}^{n}\sum_{k=0}^{n} c_{2jk}\, x^j t^k\right) - d_2\frac{\partial^2}{\partial x^2}\left(\sum_{\substack{j=0\\ j+k\le n}}^{n}\sum_{k=0}^{n} c_{2jk}\, x^j t^k\right)\right.$$

$$\left. + \left(\sum_{\substack{j=0\\ j+k\le n}}^{n}\sum_{k=0}^{n} c_{1jk}\, x^j t^k\right)^2\left(\sum_{\substack{j=0\\ j+k\le n}}^{n}\sum_{k=0}^{n} c_{2jk}\, x^j t^k\right) - b\sum_{\substack{j=0\\ j+k\le n}}^{n}\sum_{k=0}^{n} c_{1jk}\, x^j t^k\right\} dtdx = 0,$$

for q=0, 1, …, n

$$\int_{x_1}^{x_2}\int_{t_0}^{t_{q+1}}\left\{\frac{\partial}{\partial t}\left(\sum_{\substack{j=0}}^{n}\sum_{\substack{k=0\\ j+k\le n}}^{n}c_{1jk}\,x^j t^k\right)-d_1\frac{\partial^2}{\partial x^2}\left(\sum_{\substack{j=0}}^{n}\sum_{\substack{k=0\\ j+k\le n}}^{n}c_{1jk}\,x^j t^k\right)-\right. \tag{4.12}$$

$$\left.\left(\sum_{\substack{j=0}}^{n}\sum_{\substack{k=0\\ j+k\le n}}^{n}c_{1jk}\,x^j t^k\right)^2\left(\sum_{\substack{j=0}}^{n}\sum_{\substack{k=0\\ j+k\le n}}^{n}c_{2jk}\,x^j t^k\right)+(b+1)\sum_{\substack{j=0}}^{n}\sum_{\substack{k=0\\ j+k\le n}}^{n}c_{1jk}\,x^j t^k\right\}dtdx=a\int_{x_1}^{x_2}\int_{t_0}^{t_{q+1}}dtdx,$$

$$\int_{x_1}^{x_2}\int_{t_0}^{t_{q+1}}\left\{\frac{\partial}{\partial t}\left(\sum_{\substack{j=0}}^{n}\sum_{\substack{k=0\\ j+k\le n}}^{n}c_{2jk}\,x^j t^k\right)-d_2\frac{\partial^2}{\partial x^2}\left(\sum_{\substack{j=0}}^{n}\sum_{\substack{k=0\\ j+k\le n}}^{n}c_{2jk}\,x^j t^k\right)\right.$$

$$\left.+\left(\sum_{\substack{j=0}}^{n}\sum_{\substack{k=0\\ j+k\le n}}^{n}c_{1jk}\,x^j t^k\right)^2\left(\sum_{\substack{j=0}}^{n}\sum_{\substack{k=0\\ j+k\le n}}^{n}c_{2jk}\,x^j t^k\right)-b\sum_{\substack{j=0}}^{n}\sum_{\substack{k=0\\ j+k\le n}}^{n}c_{1jk}\,x^j t^k\right\}dtdx=0,$$

for q=0, 1, …, n-1

$$\vdots$$

$$\int_{x_n}^{x_{n+1}}\int_{t_0}^{t_{q+1}}\left\{\frac{\partial}{\partial t}\left(\sum_{\substack{j=0}}^{n}\sum_{\substack{k=0\\ j+k\le n}}^{n}c_{1jk}\,x^j t^k\right)-d_1\frac{\partial^2}{\partial x^2}\left(\sum_{\substack{j=0}}^{n}\sum_{\substack{k=0\\ j+k\le n}}^{n}c_{1jk}\,x^j t^k\right)-\left(\sum_{\substack{j=0}}^{n}\sum_{\substack{k=0\\ j+k\le n}}^{n}c_{1jk}\,x^j t^k\right)^2\left(\sum_{\substack{j=0}}^{n}\sum_{\substack{k=0\\ j+k\le n}}^{n}c_{2jk}\,x^j t^k\right)+\right.$$

$$\left.(b+1)\sum_{\substack{j=0}}^{n}\sum_{\substack{k=0\\ j+k\le n}}^{n}c_{1jk}\,x^j t^k\right\}dtdx=a\int_{x_n}^{x_{n+1}}\int_{t_0}^{t_{q+1}}dtdx,$$

$$\int_{x_n}^{x_{n+1}}\int_{t_0}^{t_{q+1}}\left\{\frac{\partial}{\partial t}\left(\sum_{\substack{j=0}}^{n}\sum_{\substack{k=0\\ j+k\le n}}^{n}c_{2jk}\,x^j t^k\right)-d_2\frac{\partial^2}{\partial x^2}\left(\sum_{\substack{j=0}}^{n}\sum_{\substack{k=0\\ j+k\le n}}^{n}c_{2jk}\,x^j t^k\right)\right.$$

$$\left.+\left(\sum_{\substack{j=0}}^{n}\sum_{\substack{k=0\\ j+k\le n}}^{n}c_{1jk}\,x^j t^k\right)^2\left(\sum_{\substack{j=0}}^{n}\sum_{\substack{k=0\\ j+k\le n}}^{n}c_{2jk}\,x^j t^k\right)-b\sum_{\substack{j=0}}^{n}\sum_{\substack{k=0\\ j+k\le n}}^{n}c_{1jk}\,x^j t^k\right\}dtdx=0$$

for q=0.

4.3.3 Least Square Method

From the equation (4.10) and (4.8), we get :

$$\left.\begin{aligned}
&\int_{a_1}^{b_1}\int_{a_2}^{b_2}\left\{\frac{\partial}{\partial t}\left(\sum_{\substack{j=0\\ j+k\le n}}^{n}\sum_{k=0}^{n}c_{1jk}\,x^j t^k\right)-d_1\frac{\partial^2}{\partial x^2}\left(\sum_{\substack{j=0\\ j+k\le n}}^{n}\sum_{k=0}^{n}c_{1jk}\,x^j t^k\right)-\left(\sum_{\substack{j=0\\ j+k\le n}}^{n}\sum_{k=0}^{n}c_{1jk}\,x^j t^k\right)^2\left(\sum_{\substack{j=0\\ j+k\le n}}^{n}\sum_{k=0}^{n}c_{2jk}\,x^j t^k\right)+\right.\\
&\left.(b+1)\sum_{\substack{j=0\\ j+k\le n}}^{n}\sum_{k=0}^{n}c_{1jk}\,x^j t^k\right\}\frac{\partial R_{1n}(c_{1jk},x,t)_{j+k\le n}}{\partial c_{1rq}\;{}_{r+q\le n}}\,dtdx = a\int_{a_1}^{b_1}\int_{a_2}^{b_2}\frac{\partial R_{1n}(c_{1jk}\;x,t)_{j+k\le n}}{\partial c_{1rq}\;{}_{r+q\le n}}\,dtdx\,,\\
&\int_{a_1}^{b_1}\int_{a_2}^{b_2}\left\{\frac{\partial}{\partial t}\left(\sum_{\substack{j=0\\ j+k\le n}}^{n}\sum_{k=0}^{n}c_{2jk}\,x^j t^k\right)-d_2\frac{\partial^2}{\partial x^2}\left(\sum_{\substack{j=0\\ j+k\le n}}^{n}\sum_{k=0}^{n}c_{2jk}\,x^j t^k\right)\right.\\
&\left.+\left(\sum_{\substack{j=0\\ j+k\le n}}^{n}\sum_{k=0}^{n}c_{1jk}\,x^j t^k\right)^2\left(\sum_{\substack{j=0\\ j+k\le n}}^{n}\sum_{k=0}^{n}c_{2jk}\,x^j t^k\right)-b\sum_{\substack{j=0\\ j+k\le n}}^{n}\sum_{k=0}^{n}c_{1jk}\,x^j t^k\right\}\frac{\partial R_{2n}(c_{2jk},x,t)_{j+k\le n}}{\partial c_{2rq}\;{}_{r+q\le n}}\,dtdx = 0.
\end{aligned}\right\}\quad(4.13)$$

4.3.4 Galerkin Method

From the equation (4.9) and (4.10), we get :

$$\begin{aligned}
&\int_{a_1}^{b_1}\int_{a_2}^{b_2}\left\{\frac{\partial}{\partial t}\left(\sum_{\substack{j=0\\ j+k\le n}}^{n}\sum_{k=0}^{n}c_{1jk}\,x^j t^k\right)-d_1\frac{\partial^2}{\partial x^2}\left(\sum_{\substack{j=0\\ j+k\le n}}^{n}\sum_{k=0}^{n}c_{1jk}\,x^j t^k\right)-\left(\sum_{\substack{j=0\\ j+k\le n}}^{n}\sum_{k=0}^{n}c_{1jk}\,x^j t^k\right)^2\left(\sum_{\substack{j=0\\ j+k\le n}}^{n}\sum_{k=0}^{n}c_{2jk}\,x^j t^k\right)+\right.\\
&\left.(b+1)\sum_{\substack{j=0\\ j+k\le n}}^{n}\sum_{k=0}^{n}c_{1jk}\,x^j t^k\right\}\frac{\partial P_{1n}(x,t)}{\partial c_{1rq}\;{}_{r+q\le n}}\,dtdx \qquad = a\int_{a_1}^{b_1}\int_{a_2}^{b_2}\frac{\partial P_{1n}(x,t)}{\partial c_{1rq}\;{}_{r+q\le n}}\,dtdx\,,
\end{aligned}\qquad(4.14)$$

$$\int_{a_1}^{b_1}\int_{a_2}^{b_2}\left\{\frac{\partial}{\partial t}\left(\sum_{\substack{j=0\\ j+k\le n}}^{n}\sum_{k=0}^{n} c_{2jk}\,x^j t^k\right)-d_2\frac{\partial^2}{\partial x^2}\left(\sum_{\substack{j=0\\ j+k\le n}}^{n}\sum_{k=0}^{n} c_{2jk}\,x^j t^k\right)\right.$$

$$\left.+\left(\sum_{\substack{j=0\\ j+k\le n}}^{n}\sum_{k=0}^{n} c_{1jk}\,x^j t^k\right)^2\left(\sum_{\substack{j=0\\ j+k\le n}}^{n}\sum_{k=0}^{n} c_{2jk}\,x^j t^k\right)-b\sum_{\substack{j=0\\ j+k\le n}}^{n}\sum_{k=0}^{n} c_{1jk}\,x^j t^k\right\}\frac{\partial P_{2n}(x,t)}{\partial c_{2rq}}_{r+q\le n}\,dtdx=0.$$

$$\int_{a_1}^{b_1}\int_{a_2}^{b_2}\left\{\frac{\partial}{\partial t}\left(\sum_{\substack{j=0\\ j+k\le n}}^{n}\sum_{k=0}^{n} c_{1jk}\,x^j t^k\right)-d_1\frac{\partial^2}{\partial x^2}\left(\sum_{\substack{j=0\\ j+k\le n}}^{n}\sum_{k=0}^{n} c_{1jk}\,x^j t^k\right)-\left(\sum_{\substack{j=0\\ j+k\le n}}^{n}\sum_{k=0}^{n} c_{1jk}\,x^j t^k\right)^2\left(\sum_{\substack{j=0\\ j+k\le n}}^{n}\sum_{k=0}^{n} c_{2jk}\,x^j t^k\right)+\right.$$

$$\left.(b+1)\sum_{\substack{j=0\\ j+k\le n}}^{n}\sum_{k=0}^{n} c_{1jk}\,x^j t^k\right\}x^r t^q\,dtdx \qquad = a\int_{a_1}^{b_1}\int_{a_2}^{b_2} x^r t^q\,dtdx,$$

(4.15)

$$\int_{a_1}^{b_1}\int_{a_2}^{b_2}\left\{\frac{\partial}{\partial t}\left(\sum_{\substack{j=0\\ j+k\le n}}^{n}\sum_{k=0}^{n} c_{2jk}\,x^j t^k\right)-d_2\frac{\partial^2}{\partial x^2}\left(\sum_{\substack{j=0\\ j+k\le n}}^{n}\sum_{k=0}^{n} c_{2jk}\,x^j t^k\right)\right.$$

$$\left.+\left(\sum_{\substack{j=0\\ j+k\le n}}^{n}\sum_{k=0}^{n} c_{1jk}\,x^j t^k\right)^2\left(\sum_{\substack{j=0\\ j+k\le n}}^{n}\sum_{k=0}^{n} c_{2jk}\,x^j t^k\right)-b\sum_{\substack{j=0\\ j+k\le n}}^{n}\sum_{k=0}^{n} c_{1jk}\,x^j t^k\right\}x^r t^q\,dtdx=0,$$

for i=1, 2, …, N; r,q=0, 1, …, n such that $r+q\le$n.

The equations (4.11),(4.12),(4.13),(4.14) and (4.15) are systems of L nonlinear equations in L unknowns. Solve the non-linear systems using modified Newton-Raphson method [8]. To find the unknowns and substitute the value of $\underset{j+k\le n}{c_{ijk}}$; $i=1,2$ in the equation (4.1), we get the approximate solution of (1.1).

CHAPTER FIVE

Iterative Methods

5.1 Introduction

For solving system of linear equations, we have two types of methods, which are known as "direct method and indirect method". Direct methods proceed through a finite number of steps and produce solutions that would be completely accurate. An indirect method, by contrast, produces a sequence of vectors that ideally converges to the solution.

The computations are halted when an approximate solution is obtained having some specified accuracy or after a certain number of iterations. Indirect method is always iterative in nature; a simple process is applied repeatedly to generate the sequence referred to previously iterative methods are often very efficient.

Another advantage of iterative methods is that they are usually stable, and they will actually dampen (due to round off or miner blunders) as the process continues [9], [54].

In this chapter, two iterative methods (Successive Approximation Method (SAM) [38], and Modified Successive Approximation Method (MSAM) [70]) are used to solve Brusselator system.

5.2. Successive Approximation Method [33] and [63]

The method of SAM provides a method that can, in principle, be used to solve any initial value problem

$$u' = f(t,u); \quad u(t_0) = u_0 \tag{5.1}$$

It starts by observing that any solution to (5.1) must also be a solution to

$$u(t) = u_0 + \int_{t_0}^{t} f(s, u(s))ds \tag{5.2}$$

and then iteratively constructs a sequence of solutions that get closer and closer to the actual (exact) solutions of (5.2). The SAM is based on the integral equation (5.2) as follows :

$$u_0(t) = u_0$$

$$u_1(t) = u_0 + \int_{t_o}^{t} f(s, u_0) ds$$

$$u_2(t) = u_0 + \int_{t_o}^{t} f(s, u_1(s)) ds$$

$$u_3(t) = u_0 + \int_{t_o}^{t} f(s, u_2(s)) ds$$

This process can be continued to obtain the nth approximation,

$$u_n(t) = u_0 + \int_{t_o}^{t} f(s, u_{n-1}(s)) ds, \quad n = 1,\ 2,\ \dots .$$

Then determine whether $u_n(x)$ approaches the solution $u(x)$ as n increases. This will be done by proving the following:

- The sequence $\{u_n(x)\}$ converges to a limit $u(x)$, that is $\lim_{n\to\infty} u_n(x) = u(x)$, $a \le x \le b$.
- The limiting function $u(x)$ is a solution of (5.2) on the interval $a \le x \le b$.
- The solution $u(x)$ of (5.2) is unique.

A proof of these results can be constructed along the lines of the corresponding proof for ordinary differential equations.(See [11]).

In this section we apply this method for finding an approximate solution for the Brusselator system (1.1).

5.2.1 SAM Applied to a Brusselator System

We solve the Brusselator system (1.1) by using SAM as follows:
Integrating both sides of equation (1.1) with respect to s, from 0 to t, we get:

$$\left.\begin{aligned} u(t,x) &= u(0,x) + \int_0^t \left(d_1 \frac{\partial^2 u(s,x)}{\partial x^2} - (b+1)u(s,x) + u^2(s,x)v(s,x) + a\right) ds \\ v(t,x) &= v(0,x) + \int_0^t (d_2 \frac{\partial^2 v(s,x)}{\partial x^2} + bu(s,x) - u^2(s,x)v(s,x)) ds \end{aligned}\right\} (5.3)$$

Using the boundary conditions in (1.3) we have:

$$u(t,x) = u_0(x) + \int_0^t \left(d_1 \frac{\partial^2 u(s,x)}{\partial x^2} - (b+1)u(s,x) + u^2(s,x)v(s,x) + a\right) ds$$

$$v(t,x) = v_0(x) + \int_0^t (d_2 \frac{\partial^2 v(s,x)}{\partial x^2} + bu(s,x) - u^2(s,x)v(s,x))ds \quad (5.4)$$

Start with substituting initial approximation $u_0(s,x)$ and $v_0(s,x)$ in the integral equation (5.5) to obtain a first approximation $u_1(t,x)$ and $v_1(t,x)$

$$u_1(t,x) = u_0(x) + \int_0^t (d_1 \frac{\partial^2 u_0(s,x)}{\partial x^2} - (b+1)u_0(s,x) + u_0^2(s,x)v_0(s,x) + a)ds$$

$$v_1(t,x) = v_0(x) + \int_0^t (d_2 \frac{\partial^2 v_0(s,x)}{\partial x^2} + bu_0(s,x) - u_0^2(s,x)v_0(s,x))\, ds. \quad (5.5)$$

Then this $u_1(t,x)$ and $v_1(t,x)$ is substituted again in the integral of (5.3) after replacing *t* by *s* to obtain a second approximation $u_2(t,x)$ and $v_2(t,x)$,

$$u_2(t,x) = u_0(x) + \int_0^t (d_1 \frac{\partial^2 u_1(s,x)}{\partial x^2} - (b+1)u_1(s,x) + u_1^2(s,x)v_1(s,x) + a)ds$$

$$v_2(t,x) = v_0(x) + \int_0^t (d_2 \frac{\partial^2 v_1(s,x)}{\partial x^2} + bu_1(s,x) - u_1^2(s,x)v_1(s,x))\, ds.$$

This process can be continued to obtain the nth approximation

$$u_n(t,x) = u_0(x) + \int_0^t (d_1 \frac{\partial^2 u_{n-1}(s,x)}{\partial x^2} - (b+1)u_{n-1}(s,x) + u_{n-1}^2(s,x)v_{n-1}(s,x) + a)ds$$

$$v_n(t,x) = v_0(x) + \int_0^t (d_2 \frac{\partial^2 v_{n-1}(s,x)}{\partial x^2} + bu_{n-1}(s,x) - u_{n-1}^2(s,x)v_{n-1}(s,x))\, ds$$

for n =1,2,... .

5.3 MSAM Applied to a Brusselator System

The MSAM is applied for solving Brusselator system as follows:

Rewrite equation (5.5) as follows

$$u_1(t,x) = u_0(x) + \int_0^t (d_1 \frac{\partial^2 u_0(s,x)}{\partial x^2} - (b+1)u_0(s,x) + u_0^2(s,x)v_0(s,x) + a)ds \quad (5.6)$$

$$v_1(t,x) = v_0(x) + \int_0^t (d_2 \frac{\partial^2 v_0(s,x)}{\partial x^2} + bu_1(s,x) - u_1^2(s,x)v_0(s,x))\, ds.$$

Similarly, we get

$$\left.\begin{aligned} u_2(t,x) &= u_0(x) + \int_0^t (d_1 \frac{\partial^2 u_1(s,x)}{\partial x^2} - (b+1)u_1(s,x) + u_1^2(s,x)v_1(s,x) + a)ds \\ v_2(t,x) &= v_0(x) + \int_0^t (d_2 \frac{\partial^2 v_1(s,x)}{\partial x^2} + bu_2(s,x) - u_2^2(s,x)v_1(s,x))\, ds. \end{aligned}\right\} \quad (5.7)$$

This process can be continued to obtain the nth approximation

$$u_n(t,x) = u_0(x) + \int_0^t (d_1 \frac{\partial^2 u_{n-1}(s,x)}{\partial x^2} - (b+1)u_{n-1}(s,x) + u_{n-1}^2(s,x)v_{n-1}(s,x) + a)ds$$

for n =1,2,… . (for more details on this method, see[63])

CHAPTER SIX

Numerical Examples, Conclusions and Recommendations

In this chapter, numerical examples are solved; conclusions and recommendations for future works are given.

6.1 Numerical Examples

In this section, three examples are presented and solved numerically to illustrate the efficiency of the presented method in this thesis.

Example 6.1

$$\frac{\partial u}{\partial t} = d_1 \frac{\partial^2 u}{\partial x^2} + \alpha + u^2 v - (b+1)u$$

$$\frac{\partial v}{\partial t} = d_2 \frac{\partial^2 v}{\partial x^2} - u^2 v + bu$$

$$u(0,t) = u(1,t) = a, v(0,t) = v(1,t) = \frac{b}{a}$$

$$u(x,0) = a + x(1-x), v(x,0) = \frac{b}{a} + x(1-x)$$

$$where\ a = 0.6, b = 0.2,\ d_1 = d_2 = 1/40$$

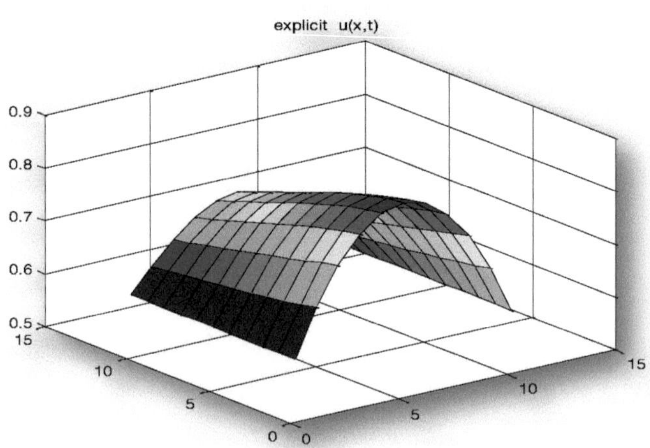

Fig. 6.1 The explicit method of Example 6.1, 0<x<1, 0<t<1

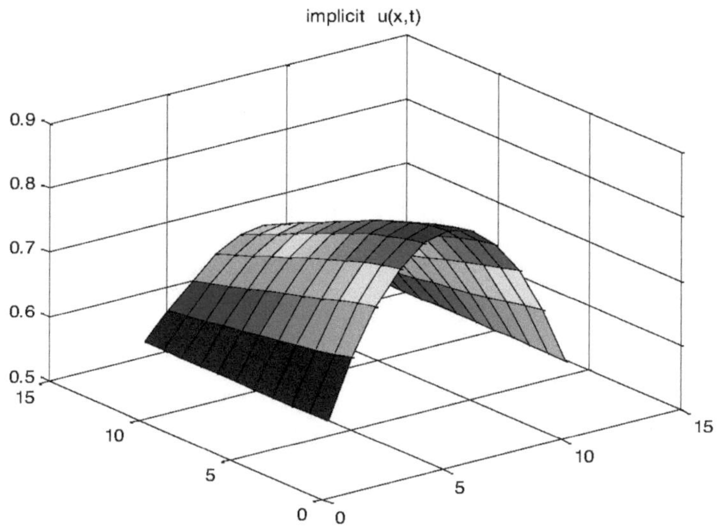

Fig. 6.2: The implicit method of Example 6.1, 0<x<1, 0<t<1

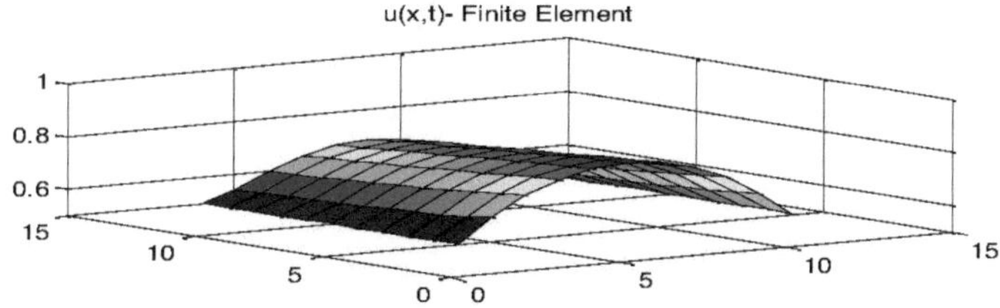

Fig. 6.3: The finite element method of Example 6.1, 0<x<1, 0<t<1

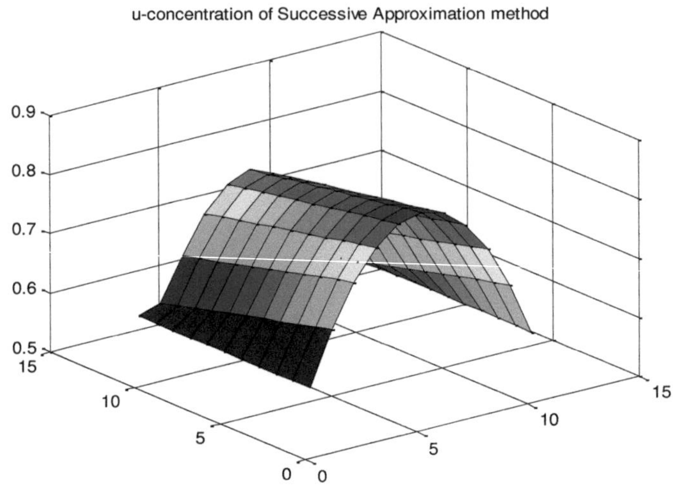

Fig. 6.4: The successive approximation method of Example 6.1, 0<x<1, 0<t<1

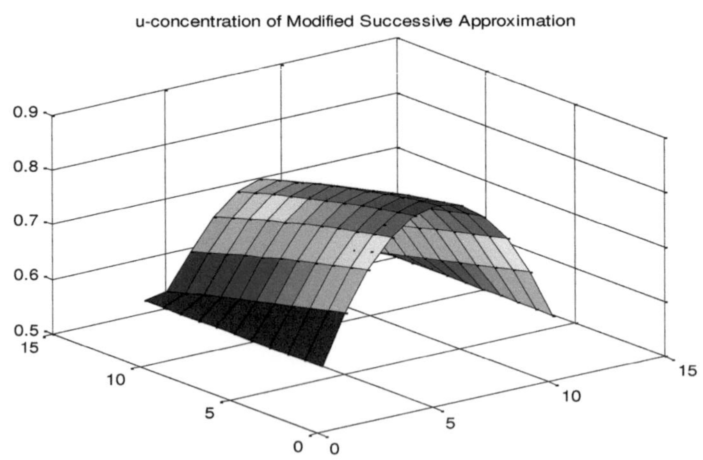

Fig. 6.5: The modified successive approximation of Example 6.1, 0<x<1, 0<t<1

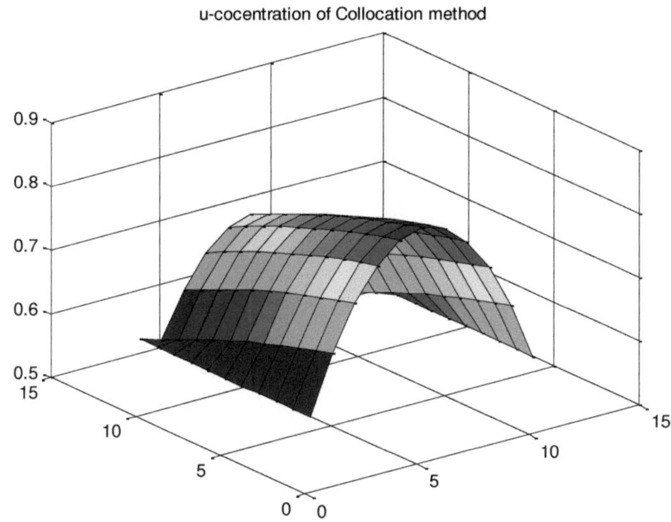

Fig. 6.6: The collocation method of Example 6.1, 0<x<1, 0<t<1

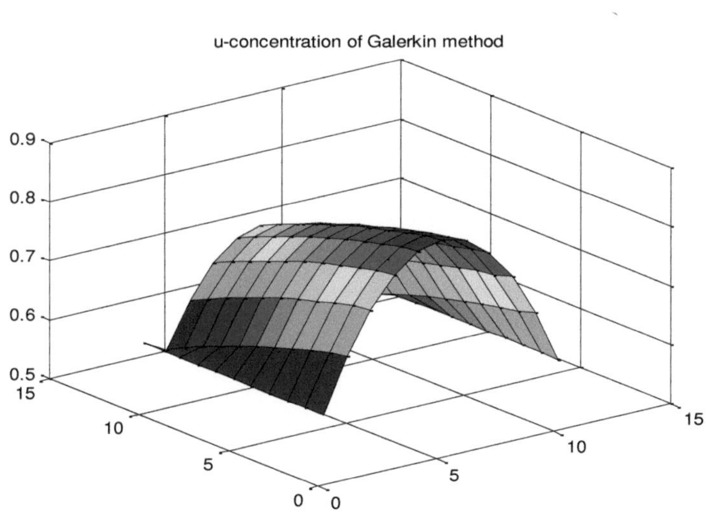

Fig. 6.7: The Galerkin method of Example 6.1, 0<x<1, 0<t<1

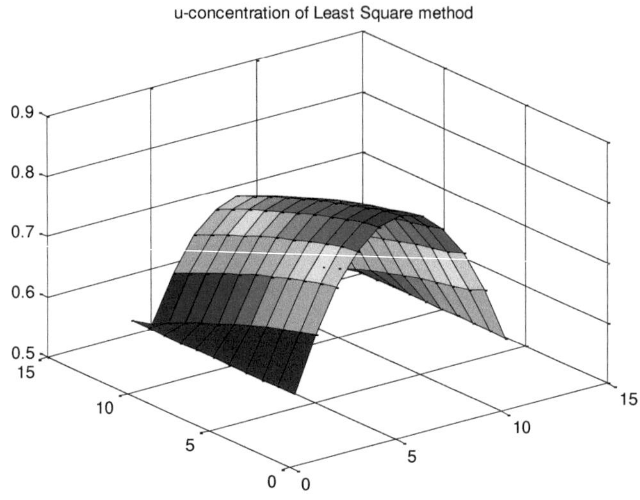

Fig. 6.8: The Least square method of Example 6.1, 0<x<1, 0<t<1

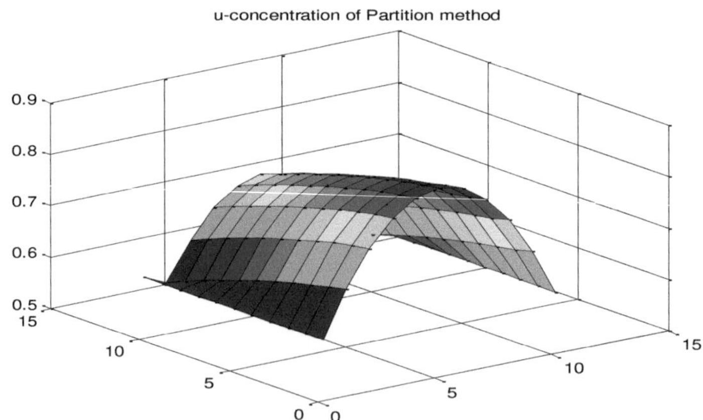

Fig. 6.9: The Partition method of Example 6.1, 0<x<1, 0<t<1

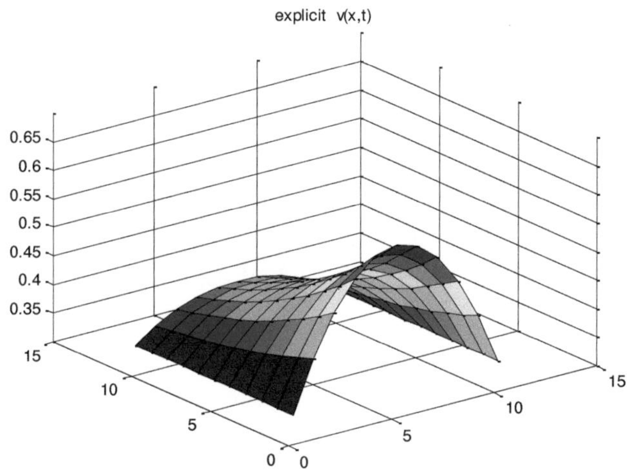

Fig. 6.10: The explicit method of Example 6.1, 0<x<1, 0<t<1

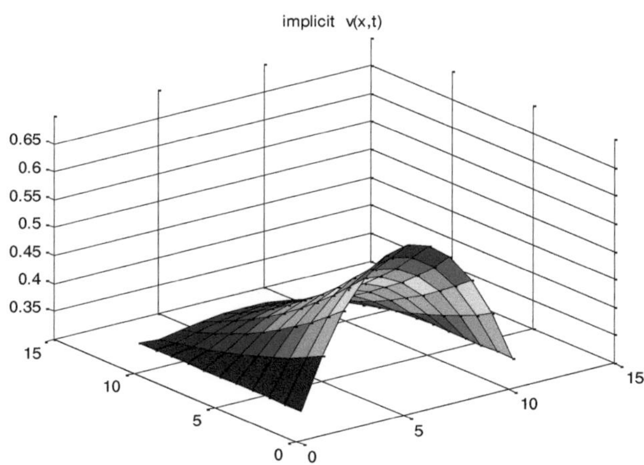

Fig. 6.11: The implicit method of Example 6.1, 0<x<1, 0<t<1

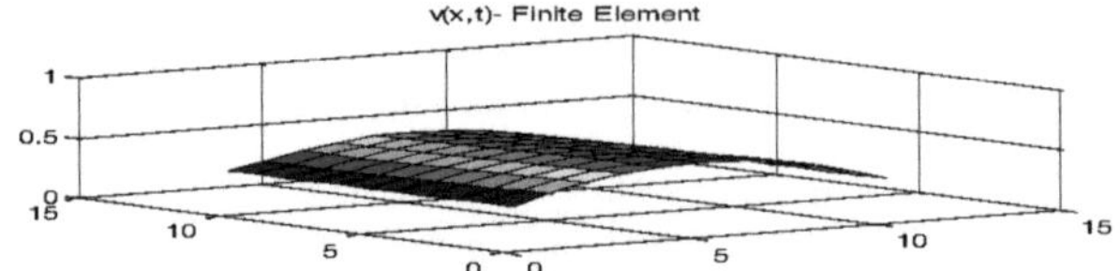

Fig. 6.12: The finite element method of Example 6.1, 0<x<1, 0<t<1

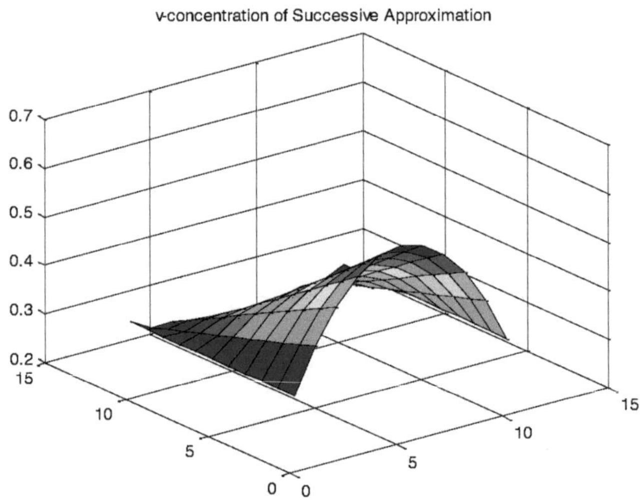

Fig. 6.13: The successive approximation method of Example 6.1, 0<x<1, 0<t<1

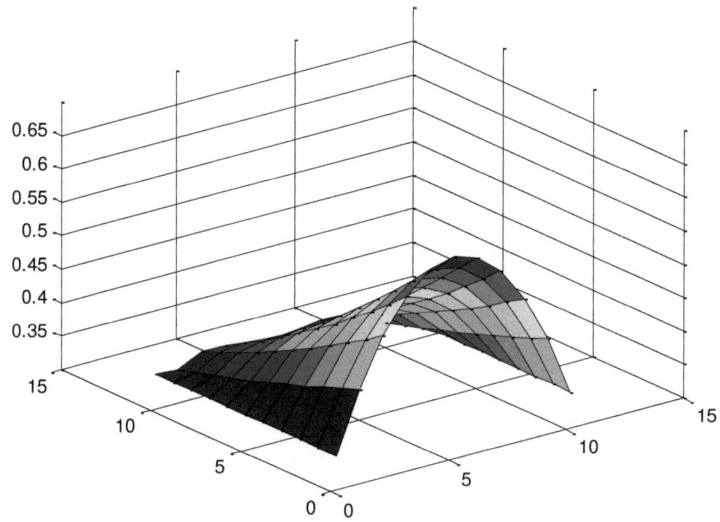

. 6.14: The modified successive approximation method of Example 6.1, 0<x<1, 0<t<1

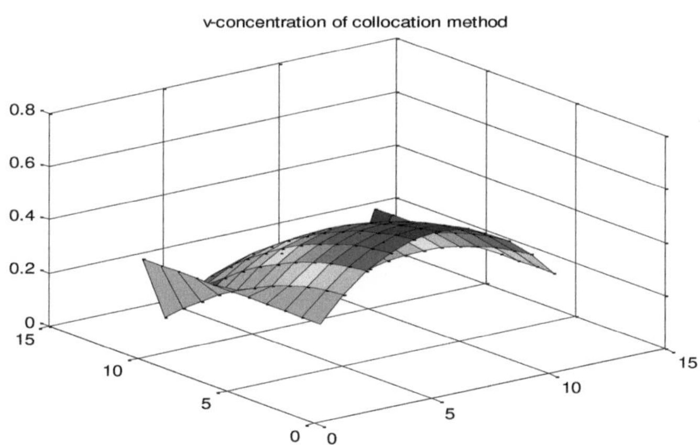

Fig. 6.15: The collocation method of Example 6.1, 0<x<1, 0<t<1

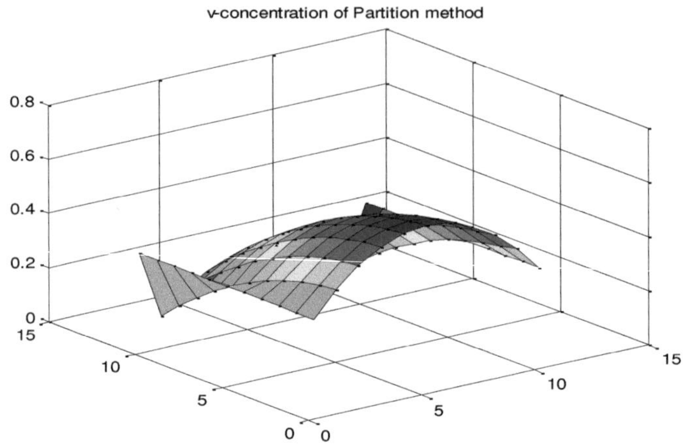

Fig. 6.16: The Partition method of Example 6.1, 0<x<1, 0<t<1

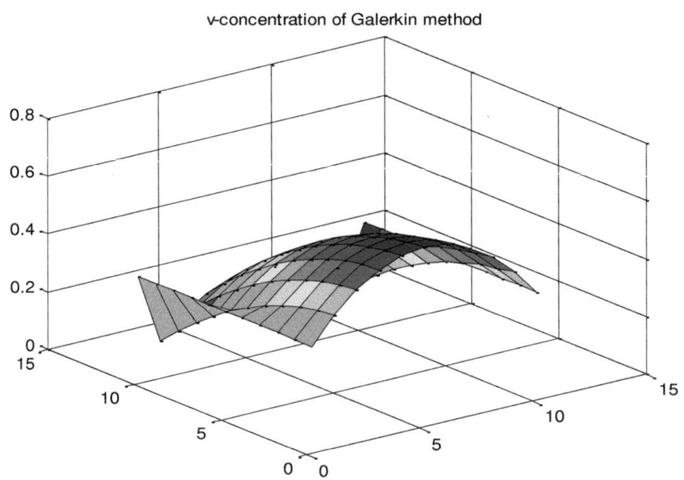

Fig. 6.17: The Galerkin method of Example 6.1, 0<x<1, 0<t<1

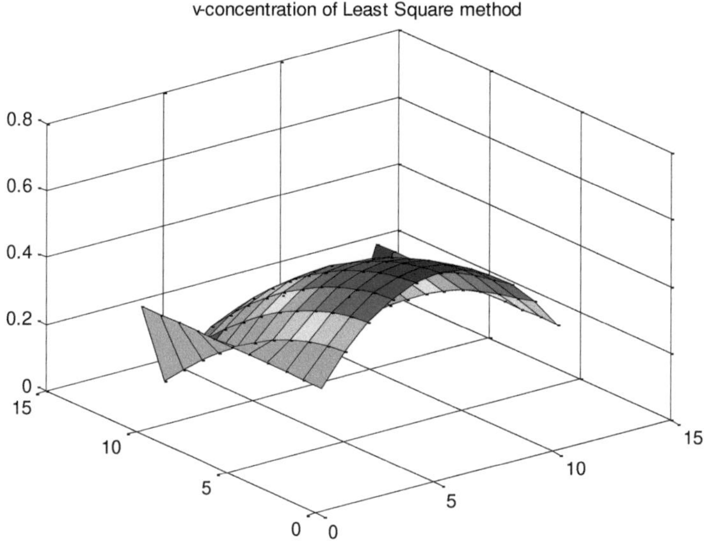

Fig. 6.18: The Least square method of Example 6.1, 0<x<1, 0<t<1

Table 6.1 Comparison between the explicit, implicit, finite element, successive approximation and modified successive approximation methods for the values of concentration u of Example 6.1

Explicit	Implicit	FEM	SAM	MSAM
0.6000	0.6000	0.6000	0.6000	0.6000
0.6628	0.6648	0.6567	0.6518	0.6505
0.7206	0.7235	0.7227	0.7215	0.7185
0.7672	0.7701	0.7756	0.7772	0.7726
0.7971	0.7999	0.8094	0.8132	0.8073
0.8074	0.8101	0.8210	0.8257	0.8193
0.7971	0.7999	0.8094	0.8132	0.8072
0.7672	0.7701	0.7757	0.7771	0.7724
0.7206	0.7235	0.7229	0.7212	0.7183
0.6628	0.6648	0.6569	0.6514	0.6502
0.6000	0.6000	0.6000	0.6000	0.6000

Table 6.2 Comparison between the collocation, Galerkin, least square, and partition methods for the concentration u(t,x) of Example 6.1

CM	GM	LSM	PM
0.6000	0.6000	0.6000	0.6000
0.6473	0.6457	0.6457	0.6457
0.7173	0.7183	0.7184	0.7179
0.7673	0.7703	0.7705	0.7698
0.7973	0.8019	0.8021	0.8012
0.8073	0.8129	0.8131	0.8121
0.7973	0.8033	0.8036	0.8026
0.7673	0.7732	0.7735	0.7726
0.7173	0.7226	0.7228	0.7222
0.6473	0.6515	0.6516	0.6513
0.6000	0.6000	0.6000	0.6000

Table 6.3 Comparison between the explicit, implicit, finite element, successive approximation and modified successive approximation methods for the concentration v(t,x) of Example 6.1

Explicit	Implicit	FEM	SAM	MSAM
0.3333	0.3333	0.3333	0.3333	0.3333
0.3806	0.3964	0.3938	0.3665	0.3716
0.4186	0.4375	0.4647	0.4018	0.4111
0.4444	0.4621	0.5170	0.4211	0.4332
0.4585	0.4747	0.5481	0.4300	0.4436
0.4630	0.4786	0.5583	0.4326	0.4466
0.4585	0.4747	0.5482	0.4301	0.4437
0.4444	0.4621	0.5175	0.4212	0.4333
0.4186	0.4375	0.4665	0.4020	0.4113
0.3806	0.3964	0.3981	0.3668	0.3718
0.3333	0.3333	0.3333	0.3333	0.3333

Table 6.4 Comparison between the collocation, Galerkin, least square, and partition methods for the concentration v(t,x) of Example 6.1

CM	PM	GM	LSM
0.3333	0.3333	0.3333	0.3333
0.3277	0.3256	0.3277	0.3266
0.3977	0.3979	0.4002	0.3992
0.4477	0.4497	0.4522	0.4512
0.4777	0.4811	0.4837	0.4828
0.4877	0.4920	0.4947	0.4938
0.4777	0.4825	0.4852	0.4842
0.4477	0.4525	0.4551	0.4542
0.3977	0.4021	0.4046	0.4036
0.3277	0.3312	0.3335	0.3325
0.3333	0.3333	0.3333	0.3333

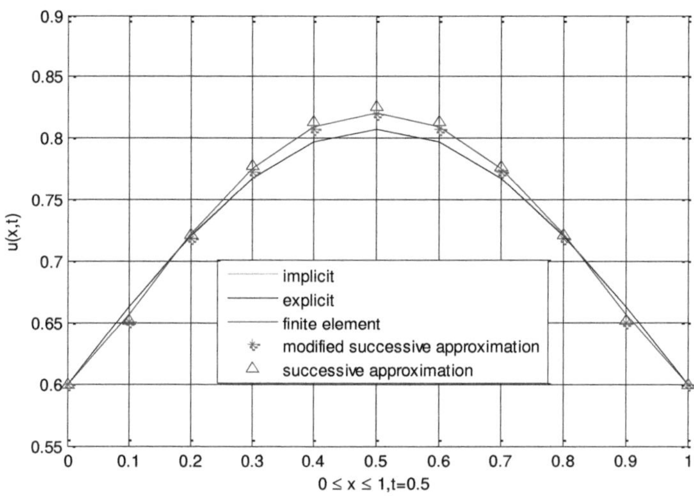

Fig. 6.19 Comparison between explicit, implicit, finite element, successive approximation and modified successive approximation methods

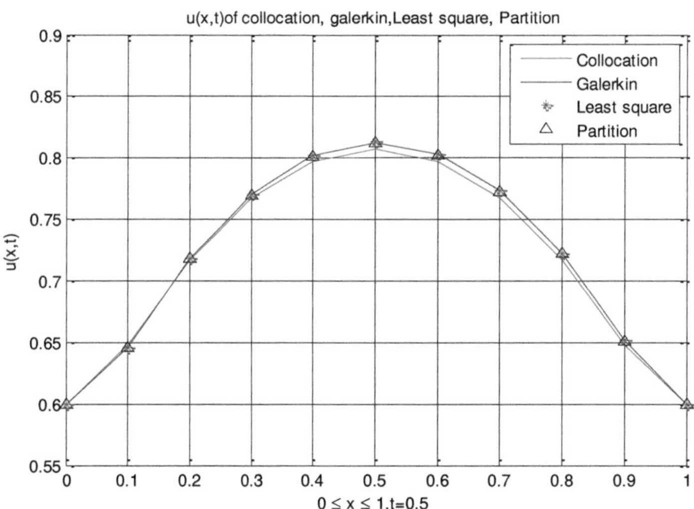

Fig. 6.20 Comparison between collocation,Galerkin,Least square and Partition methods

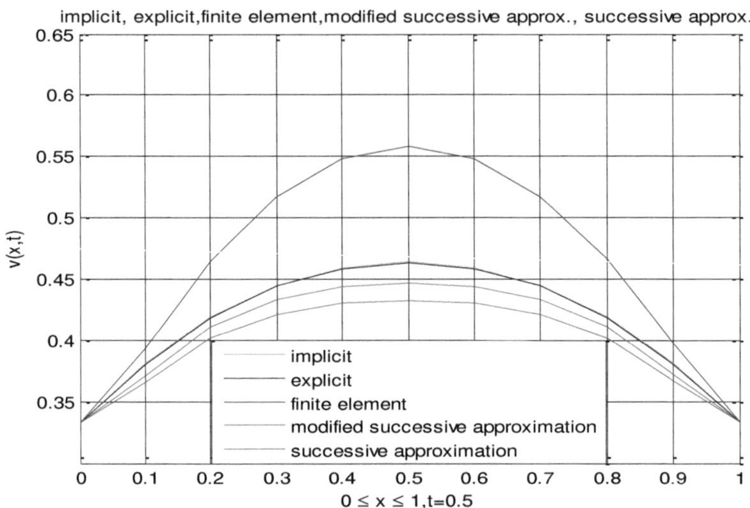

Fig . 6.21 Comparison between implicit, explicit, finite element, modified successive approximation and successive approximation.

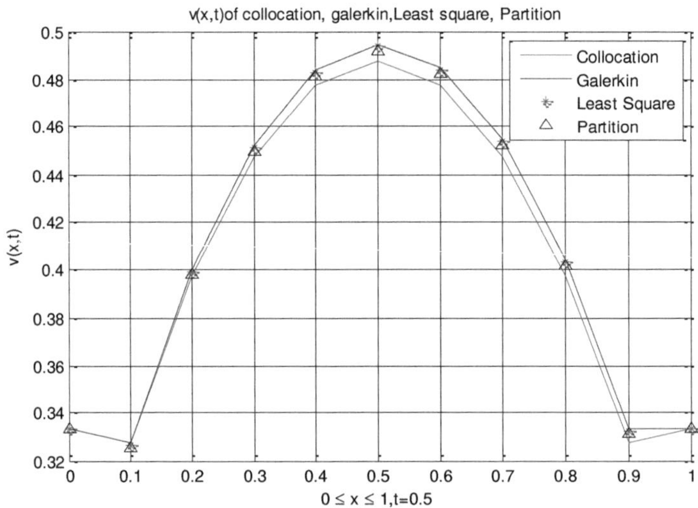

Fig. 6.22 Comparison between collocation, Galerkin, Least square and Partition mehods

Table 6.5 Comparison between explicit, implicit, finite element, successive approximation, modified successive approximation, collocation, Galerkin, least square, partition methods of v-concentration of Example 6.1

Explicit	Implicit	FEM	MSAM	SAM	CM	GM	LSM	PM
0.3333	0.3333	0.3333	0.3333	0.3333	0.3333	0.3333	0.3333	0.3333
0.3806	0.3964	0.3938	0.3716	0.3665	0.3277	0.3277	0.3266	0.3256
0.4186	0.4375	0.4647	0.4111	0.4018	0.3977	0.4002	0.3992	0.3979
0.4444	0.4621	0.5170	0.4332	0.4211	0.4477	0.4522	0.4512	0.4497
0.4585	0.4747	0.5481	0.4436	0.4300	0.4777	0.4837	0.4828	0.4811
0.4630	0.4786	0.5583	0.4466	0.4326	0.4877	0.4947	0.4938	0.4920
0.4585	0.4747	0.5482	0.4437	0.4301	0.4777	0.4852	0.4842	0.4825
0.4444	0.4621	0.5175	0.4333	0.4212	0.4477	0.4551	0.4542	0.4525
0.4186	0.4375	0.4665	0.4113	0.4020	0.3977	0.4046	0.4036	0.4021
0.3806	0.3964	0.3981	0.3718	0.3668	0.3277	0.3335	0.3325	0.3312
0.3333	0.3333	0.3333	0.3333	0.3333	0.3333	0.3333	0.3333	0.3333

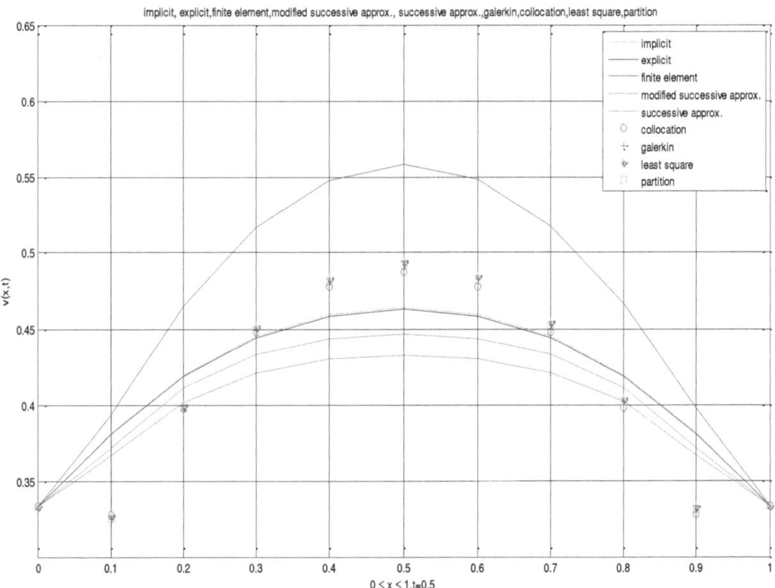

Fig. 6.23 Comparison between implicit, explicit, finite element, modified successive approximation, successive approximation, collocation, Galerkin, Least square and partition methods for v-concentration of Example 6.1.

Table 6.6 Comparison between explicit, implicit, finite element, successive approximation, modified successive approximation, collocation, Galerkin, least square and partition methods of u-concentration of Example 6.1

Explicit	Implicit	FEM	MSAM	SAM	CM	GM	LSM	PM
0.6000	0.6000	0.6000	0.6000	0.6000	0.6000	0.6000	0.6000	0.6000
0.6628	0.6648	0.6567	0.6505	0.6518	0.6473	0.6457	0.6457	0.6457
0.7206	0.7235	0.7227	0.7185	0.7215	0.7173	0.7183	0.7184	0.7179
0.7672	0.7701	0.7756	0.7726	0.7772	0.7673	0.7703	0.7705	0.7698
0.7971	0.7999	0.8094	0.8073	0.8132	0.7973	0.8019	0.8021	0.8012
0.8074	0.8101	0.8210	0.8193	0.8257	0.8073	0.8129	0.8131	0.8121
0.7971	0.7999	0.8094	0.8072	0.8132	0.7973	0.8033	0.8036	0.8026
0.7672	0.7701	0.7757	0.7724	0.7771	0.7673	0.7732	0.7735	0.7726
0.7206	0.7235	0.7229	0.7183	0.7212	0.7173	0.7226	0.7228	0.7222
0.6628	0.6648	0.6569	0.6502	0.6514	0.6473	0.6515	0.6516	0.6513
0.6000	0.6000	0.6000	0.6000	0.6000	0.6000	0.6000	0.6000	0.6000

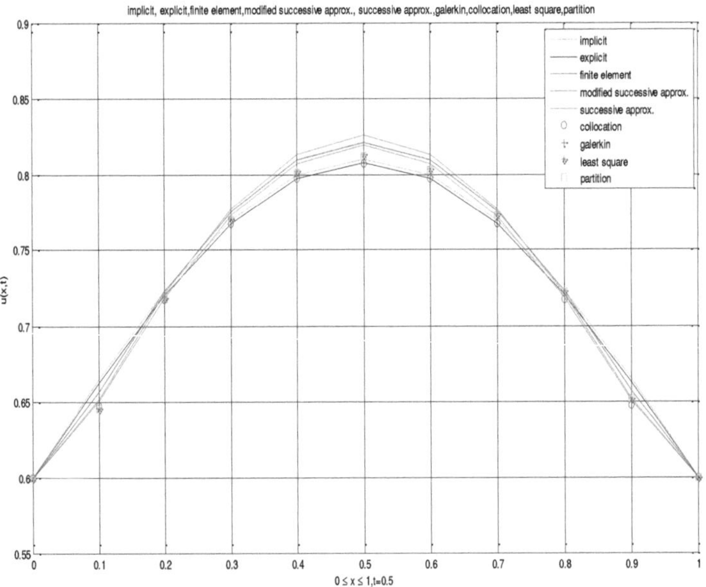

Fig. 6.24 Comparison between implicit, explicit, finite element, modified successive approximation, successive approximation, collocation, Galerkin, Least square and partition methods for u-concentration of Example 6.1.

Example 6.2

$$\frac{\partial u}{\partial t}=\frac{1}{4}\left(\frac{\partial^2 u}{\partial x^2}+\frac{\partial^2 u}{\partial y^2}\right)+u^2v-2u$$

$$\frac{\partial v}{\partial t}=\frac{1}{4}\left(\frac{\partial^2 v}{\partial x^2}+\frac{\partial^2 v}{\partial y^2}\right)-u^2v+u$$

in the region $R=\{(x,y): x^2+y^2<1,\ x>0,\ y>0\}$
subject to the initial -boundary conditions
$(u(x,y,0),v(x,y,0))=(\exp(-x-y),\exp(x+y)$, *for* $(x,y)\in R$,
$(u(0,y,t),v(0,y,t))=(\exp([-t/2-y],\exp[t/2+y])$, *for* $0<y<1,\ t>0$,
$(u(x,0,t),v(x,0,t))=(\exp([-t/2-x],\exp[t/2+x])$, *for* $0<x<1,\ t>0$,
$(\frac{\partial u}{\partial n},\frac{\partial v}{\partial n})=[x+y]\times(-\exp[-t/2-x-y],\exp[t/2+x+y])$, *for* $x^2+y^2=1,\ t>0$.
$(u,v)=(\exp[-t/2-x-y],\ \exp[\ t/2+x+y])$, is the exact solution of the problem.

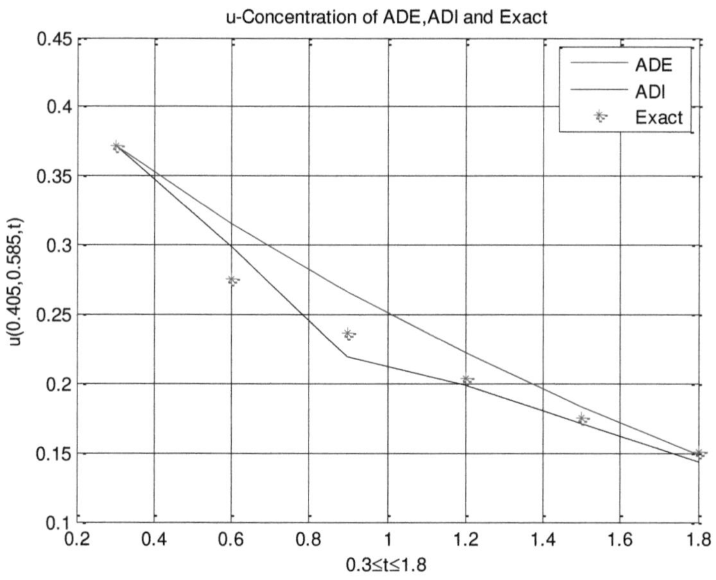

Fig. 6.25 Comparison between ADE, ADI and Exact solution of Example 6.2

Table 6.7 comparison between the methods ADE , ADI, and the exact solution for the values of concentrations *u*.

ADE	ADI	Exact Solution
0.371576691022046	0.371576691022046	0.371576691022046
0.315849593538622	0.298639421159804	0.275270783089752
0.266352962586403	0.219563521894050	0.236927758682122
0.222299636371069	0.199079403982470	0.203925611734213
0.183105898844905	0.171007857085359	0.175520400616997
0.148401742114595	0.143371805026434	0.151071808836371

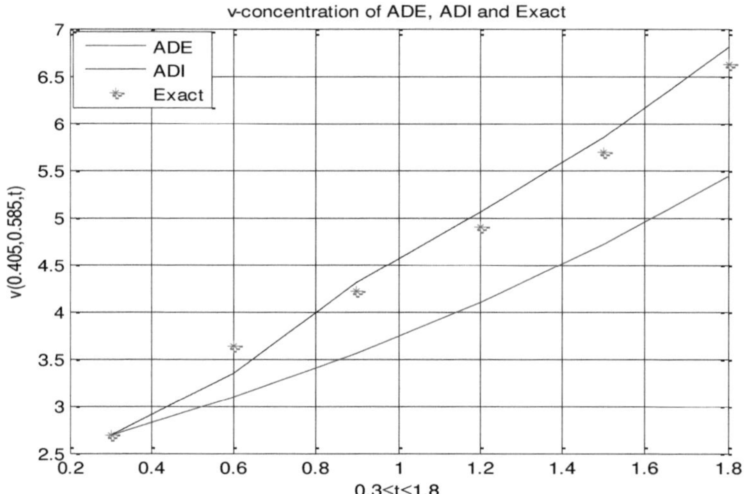

Fig. 6.26 comparison between ADE, ADI and Exact solution of Example 6.2

Table 6.8 Shows the comparison between the methods ADE , ADI, and the exact solution for the values of concentrations *v*.

ADE	ADI	Exact Solution
2.691234472349263	2.691234472349263	2.691234472349263
3.094987769672571	3.349347686080280	3.632786555752809
3.561441469711582	4.310824337296547	4.220695816996552
4.099854965409280	5.065523044067986	4.903748928326623
4.720792386658257	5.848409505303884	5.697343422671991
5.436027130989021	6.810472684422297	6.619368681043077

Example 6.3:

$$\frac{\partial u}{\partial t}=\frac{1}{500}\left(\frac{\partial^2 u}{\partial x^2}+\frac{\partial^2 u}{\partial y^2}\right)+u^2 v-\frac{3}{2}u+1$$

$$\frac{\partial v}{\partial t}=\frac{1}{500}\left(\frac{\partial^2 v}{\partial x^2}+\frac{\partial^2 v}{\partial y^2}\right)-u^2 v+\frac{1}{2}u$$

in the region $R=\{(x,y):0<x<1,\ 0<y<1\}$

subject to

$$(u(x,y,0),v(x,y,0))=\left(\frac{1}{2}x^2-\frac{1}{3}x^3,\frac{1}{2}y^2-\frac{1}{3}y^3\right), \text{for } (x,y)\in R,$$

$\left(\frac{\partial u}{\partial x},\frac{\partial v}{\partial x}\right)=(0,0)$ at the point $(0,y)$ and $(1,y)$ for $0<y<1$, and $t>0$..

$\left(\frac{\partial u}{\partial y},\frac{\partial v}{\partial y}\right)=(0,0)$ at the point $(x,0)$ and $(x,1)$ for $0<x<1$, and $t>0$.

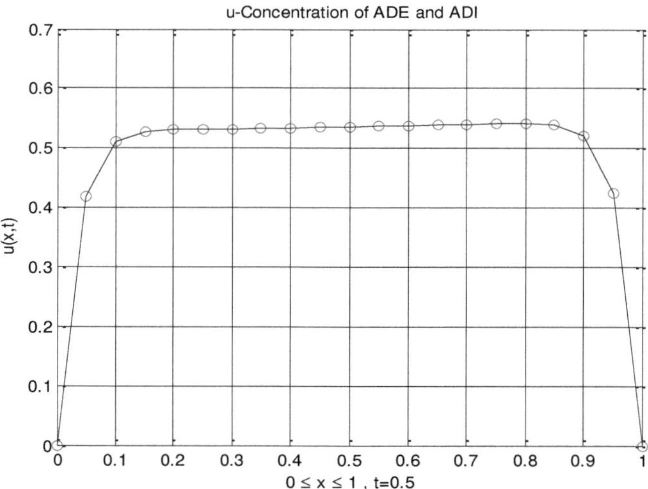

Fig. 6.27 Comparison between ADE and ADI methods for Example 6.3

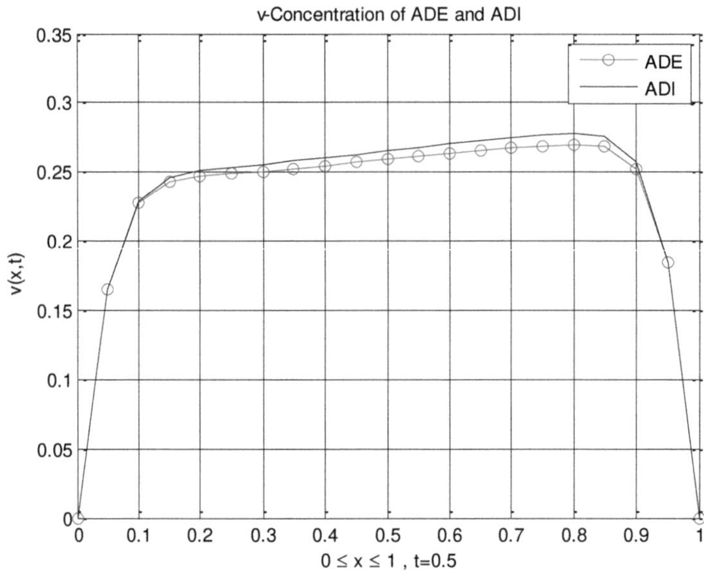

Fig. 6.28 Comparison between ADE and ADI methods for Example 6.3

Table 6.9 Comparison between ADE and ADI for both u-concentration and v-concentration of Example 6.3

ADEu	ADIu	ADEv	ADIv
0	0	0	0
0.418093427313441	0.418037301762852	0.165257693128643	0.165016678974309
0.510414731833183	0.510362471583337	0.227209212158516	0.228796399787785
0.527375346605206	0.527325579062866	0.243023394200850	0.246377770102526
0.530290619712331	0.530244981995636	0.246880920830135	0.251217253956370
0.531235609321934	0.531196663908266	0.248804566125793	0.253596133094942
0.532042008549831	0.532011233994595	0.250665524729720	0.255751908754085
0.532909934507732	0.532888313934932	0.252669854367180	0.258039101501989
0.533840330420088	0.533828608772390	0.254792161892842	0.260465503662110
0.534815171860345	0.534813893969458	0.256984996231137	0.262985487482971
0.535814753103609	0.535824259802292	0.259200760006308	0.265546801433325
0.536818935466600	0.536839354213003	0.261393537313655	0.268097126687050
0.537807201362198	0.537838435999103	0.263519270965989	0.270584804845593
0.538758580265913	0.538800301048557	0.265535652679890	0.272957944535160
0.539650231184560	0.539701869063372	0.267400165287487	0.275156669925136
0.540437826720320	0.540498592267962	0.269045533328307	0.277057905618244
0.540851182134286	0.540920056156999	0.270144142564714	0.278130625273590
0.538792619824467	0.538866415516789	0.268496736593505	0.275640264393692
0.521055351506129	0.521113469332205	0.252760380553536	0.257508282109768
0.423988200093128	0.423976735507793	0.184508844827995	0.185048452592917
0	0	0	0

Discussion of Tables and Figures

It is seen from tables and figures for one dimensional Brusselator model in Example 6.1 that all the methods are close to each other, as shown in the Figures 6.19, 6.20, 6,21 and 6.22, and Tables 6.1, 6.2, 6.3 and 6.4. But it is clear that finite element method is more accurate than other methods (MSAM, SAM, FDM, WRM) as shown in the Figures 6.23 and 6.24, and tables 6.5 and 6.6. Also for finite difference methods in one dimension, the Crank – Nicolson (CN) method is more accurate than explicit method as shown in Figures 6.19 and 6.20, and Tables 6.1and 6.3.

In Example 6.2 for two dimension Brusselator model with exact solution, it is clear that ADI method is much more close to the exact solution than ADE method, see Figures.6.25 and 6.26, and Tables 6.5 and 6.6.

In Example 6.3, for two dimensional Brusselator model without exact solution, ADI method is more accurate than ADE method as shown in the Figure 6.27 and 7.28 and Table 6.9.

6.2 Conclusions

We solved the Brusselator model in one dimension using FDM (explicit and implicit methods), FEM, SAM, MSAM, WRM including (CM, GM, LSM and PM) and we found that for the FDM, the implicit method is more accurate than the explicit method, but the explicit finite difference method is faster than the implicit finite difference method as shown in the Figures 6.19 and 6.21, and Tables 6.1 and 6.3, and for weighted residual methods we found that all methods give similar results, but CM is simpler and faster than the other methods as illustrated in the Figures 6.20 and 6.22, and Tables 6.2 and 6.4.

For comparing all the methods, we found that FEM is more accurate than the other methods MSAM, SAM, FDM (explicit and implicit methods) and WRM as shown in the Figures 6.23 and 6.24.

Also we studied the Brusselator model in two dimension using Alternating Direction Explicit method (ADE) and Alternating Direction Implicit method (ADI) and we found that Alternating Direction Implicit method is more accurate than Alternating Direction Explicit method as shown in the Figures 6.25, 6.26, 6.27 and 6.28, and Tables 6.7 ,6.8 and 6.9.

We also studied the numerical stability for FDM (explicit and implicit methods) in one dimension and we found that explicit method is conditionally stable , while the implicit method is unconditionally stable, and for two dimensional space we found that ADE method is conditionally stable while ADI method is unconditionally stable .

6.3 Recommendations

1) Studying the bifurcation analysis of Brusselator model.

2) Solving the Brusselator model by (Adomian decomposition method), (using different types of spline functions) and (power series and pade series) with studying the stability analysis of the

Brusselator model

REFERENCES

[1] Abdulgafor Mohammad Amin Al-Rozbayani, On the Numerical Solution of System of Nonlinear Parabolic Partial Differential Equations, *University of Mosul, Ph.D Thesis,2006.*

[2] Alkis, C., Numerical Methods for Chemical Engineers with Matlab Applications, *New Jersey, 1999.*

[3] Allaire, P. E., Basic of the Finite Element Method, *Wn. C. Brown publishers, 1985.*

[4] Ames, W. F., Numerical Methods for Partial Differential Equations. *3rd edition. Academic, Inc. 1992.*

[5] Auchmuty, J. F. G. and Nicolis, G., Bifurcation Analysis of Nonlinear Reaction-Diffusion Equations-I. *Bull. Math. Biol., Vol.37 (1975),* 1-43.

[6] Brauer, F. and Nohel, J. A., Ordinary Differential Equations: a First Course, *2nd edition, W. A. Benjamin, INC. London, 1973.*

[7] Castets, V.; Dulos, E.; Biossonade, J. and De Kepper, P., Experimental evidence of a sustained standing Turing-type nonequilibrium chemical pattern, *Phys. Rev.Lett.Vol.64,No. 2953(1990).*

[8] Chapra, S. C. and Canale, R. R., Numerical Methods for Engineers: With Software and Programming Application, *4th Edition, McGraw-Hill, 2002.*

[9] Charkravarti, S.; Marek, M. and Ray, W. H.,Reaction-diffusion system with Brusselator kinetics:Control of a quasiperiodic route to chaos., *Phys. Rev. E. Vol.52, No.2407(1995) .*

[10] Charm, S. E., Fundamentals of Food Engineering. *Westport, CN: The Avi Publication Company, 1973.*

[11] Coddington, E. A., An Introduction to Ordinary Differential Equations, *Prentice-Hall, INC., 1961.*

[12] Crank, J., Mathematics of Diffusion, *2nd edition. Oxford: Oxford University Press, 1975.*

[13] Cross, M. C. and Hohenberg, P. C.,Pattern formation outside of equilibrium,

Rev. Mod. Phys. Vol.65, No.851 (1993).

[14] Debnath, L., Nonlinear Partial Differential Equations for Scientists and Engineers, *Birkhauser Boston, 1997.*

[15] Decpak, V.; Dimitrios, V. and Domiel, A., A discontinuous Galerkin formulation for solution of parabolic equations on nonconforming meshes, *(2005), www.cims.nyu.edu/dd16/proceeding/kulkarni contrib. pdf.*

[16] De Kepper, P.; Castets, V.; Dulos, E. and Biossanade,Turing –type chemical patterns in the chlorite-iodide-malonic acid reaction, J., *Physica D. Vol.49, No.161(1991*).

[17] Desai, S. C. and Abel, J. F., Introduction to the finite element, *Van Nostrand, New York 1972.*

[18] DeWel G.; Bachir, M.; Metens, S. and Borckmans, P., Faraday Discuss. *Vol.120, No.363 (2001).*

[19] Dick, D.A .T., The rate of Diffusion of Water in the Protoplasm of Living Cells, *Expl. Cell Res., Vol.17, 5-13 ,(1959)*

[20] Ebrahim, A.; Kurosh, S.; Mousa, F. and Ibrahimi K., Analytical approximate solution of the cooling problem by A domain decomposition method, *Communications in Nonlinear Science and Numerical Simulation*, *Vol.14, (2009), 462-472.*

[21] Ellbeck, J. C., The Pseude-Spectral Method and Path Following Reaction–Diffusion Bifurcation Studies, *SIAM J. Sci. Stat. Comput., Vol. 7, No.2(1986), 599-610.*

[22] Erneux,T. and Herschkowitz-Kaufmann, M., Rotating Waves as Asymptotic Solutions of a Model Chemical Reaction, *J. Chem. Phys. Vol.66(1977), 248-250.*

[23] Guckenheimer, J., *In* Dynamical systems and Turbulence, war-wick, *Lecture Notes in Mathematics, Vol. 898 (1981), 99-142.*

[24] Guymon, G. L., A finite element solution of the one dimensional diffusion-convection equation, *water resources research, Vol.6, No.1 (1970), 204-210.*

[25] Hassani, S., Mathematical methods: for students of Physics and related fields,

Springer, 2000.

[26] Hays, D. F. and Curd, H. N., Concentration Dependent Diffusion in a Semi infinite Medium, *J. Franklin Inst., Vol.283 (1967), 300-308*

[27] Herschkowitz- Kaufman, M., Bifurcation Analysis of Nonlinear Reaction Diffusion Equations—II. Steady State Solutions and Comparison With Numerical Simulation, *Bull. Math. Bio., Vol.37(1975), 589-636.*

[28] Hlavacek,V. and Rompay, P. V., Current Problems of Multiplicity, Stability and Sensitivity of States in Chemically Reacting Systems, *Chem. Engng. Sci., Vol.36 (1981), 1587-1597.*

[29] Holloway, D. M. and Harrison, L. G., Order and localization in reaction-diffusion pattern, *Physica A, Vol.222 (1995), 210-233.*

[30] Hrsschkowitz-Kaufman, M. and Erneux T., *Ann. Acad. Sci., Vol.78, No.296 (1979).*

[31] Hu, J. W. and Tian, C. S., A Galerkin Partial Differential upwind Finite Element Method for the convection diffusion equations, *Chinese Series, (1992), 446-459.*

[32] Jain, M. K., Numerical solution of differential equations, *John Wiley & Sons, New York, 1979.*

[33] Jerri, A. J., Introduction to integral equations with applications, *Marcel Dekker, Inc, New York and Basel, 1985.*

[34] Jost. J., Partial differential equations, *Springer-Verlag, New York, 2002.*

[35] Kang, H. and Pesin, Y., Dynamics of a Discrete Brusselator Model: Escape to Infinity and Julia Set, *Milan J. Math., Vol.73 (2005)1-17.*

[36] Karafyllis, I.; Chrisofides, P. D. and Daoutidis, P., Dynamics of a reaction diffusion system with Brusselator kinetics under feedback control, *Vol.59, No.1 (1999), 372-379.*

[37] Kubicek, M.; Ryzler, V. and Marek, M., Spatial Structures in a Reaction-Diffusion System—Detailed Analysis of the Brusselator, *Biophys. Chem., Vol.8 (1978), 235-246*

[38] Kuptsov, P. V.; Kuznetsov, S. P. and Mosekilde, E., Particle in the Brusselator

model with flow, *Physica D, Vol.163(2002),80-88.*

[39] Lapidus, L, and G. F., Numerical Solution of Partial Differential Equations in Science and Engineering, *John Wiley &Sons Inc, 1982.*

[40] Lefever, R. and Prigogine, I., Symmetry Breaking Instabilities in Dissipative Systems II, *J. Chem. Phys. Vol.48 (1968), 1695-1700.*

[41] Leppanen, T.; Karttunen, M.; Kaski, K.; Barrio, R. A. and Zhang, L.
A new dimension to Turing patterns, *Physica D, Vol.168, No.32(2002).*

[42] Leppanen, T.; Karttunen, M.; Barrio, R. A. and Kaski, K.,Morphological transitions and bistability in Turing systems, *Phys. Rev. E, Vol.70, No.066202 (2004).*

[43] Le Veque, R. J., Finite Difference Method for Differential Equations, *University of Washington, 2005.*

[44] Li, R. S.and Nicolis, G., Bifurcation Phenomena in Nonideal Systems.2. Effect of Nonideality correction on Symmetry-Breaking Transitions, *J. Phys. Chem. Vol.85 (1981), 1912-1918.*

[45] Lin, S. H., Nonlinear Diffusion in Biological Systems, *Bull. math. Biol. Vol.41 (1979), 151-162.*

[46] Logan, J. D., Applied Mathematics, *John Wiley &Sons, 1987.*

[47] Manaa, S.A and Hassin, N. A., Stability Analysis of Steady State Solutions of Sine-Gordon Equation, *Raf. J. Of Comp. & Maths., Vol.3, No.2 (2006).*

[48] Mathews, J. H. and Fink, K. D., Numerical methods using Matlab, *Prentice-Hall, Inc., 1999.*

[49] Matzinger, E., Asymptotic behavior of solutions near a turning point: The example of the Brusselator equation, *Journal of Differential Equations, Vol.220 (2006), 478-510.*

[50] Murray, J. D., Mathematical Biology, 2nd *edition, Springer-Verlag, Berlin, 1993.*

[51] Nandapurkar, P. and Hlavacek, V., Concentration-dependent diffusion coefficient and dissipative structures,*J. Bulletin of Mathematical Biology, Vol. 46 , 2(1984), 269-282 .*

[52] Nicolis,G.; Erneux, T. and Herschkowitz-Kaufman, M., Pattern Formation in

Reacting and Diffusing Systems. *Avd. Chem. Phys. Vol.38 (1978), 263-315.*

[53] Nicolis, G. and Prigogine, I., Self-Organization in Nonequilibrium Systems, *New York:John Wiley, 1977.*

[54] Otto, S. R. and Denier, J. P., An Introduction to Programming and Numerical Methods in MATLAB, *Springer, 2005.*

[55] Pastushenko, V. P. and Schindler, H., Distance of the reduced Brusselator from equilibrium Steady-State and relaxation regimes, *Eur Biophy. J., Vol.27(1998),227-236*

[56] Peaceman, D. W. and Rachford, H. H., The Numerical Solution of Parabolic and Elliptic Different Equations, *SIAM J.*, Vol.3 (1955), pp.28-41.

[57] Petrov, V.; Metens, S.; Borckmans, P.; Dewel, G. and Show-alter, K.,*Tracking* Unstable Turing Pattern through Mixed-Mode Spatiotemporal Chaos, *Phys. Rev. Lett.*, Vol.75, No.2895 (1995).

[58] Qin, F., Wolf, E. E. and Kevrekidis, H.-C., *AIChE. J., Vol.44, No.1579 (1998).*

[59] Quarteroni, A. and Valli, A., Numerical Approximation of Partial Differential Equations, *Springer-Verlag, 1997.*

[60] Quyang, Q. and Swinney, H. L., *Chaos*, Vol.1, No.441 (1991) .

[61] Quyang, Q. and Swinney, H. L., *Nature(London)*, Vol.352, No.610 (1991).

[62] Richard, G. R. and Duong, D. D., Applied Mathematics and Modeling for Chemical engineers, *John Wiley & Son, Inc., 1995.*

[63] Saeed, R. K., Computational methods for solving system of linear Volterra and integro-differential equations, *Ph.D. Thesis, Salahaddin University/Erbil, Iraq, 2006.*

[64] Satnoianu, R. A.; Merkin, J. H. and Scott, S. K., Pattern formation in a differential-flow reactor model, *Chemical Engineering Science*, Vol.55(2000), 461-469

[65] Schnakenberg, J., Simple Chemical Reaction System with Limit Cycles Behaviour, *J. Theo. Biol.*, Vol.81 (1979), 389-400.

[66] Shoji, H.; Yamada, K.; Ueyama, D. and Ohta, T. Turing Patterns in three dimensions, *PHYSICAL REVIEW E*, Vol.75, No.046212 (2007)(13 pages)..

[67] Shanthakumar, M., Computer Based Numerical Analysis, *Khanna publishers, Neisaraic Delhi-110006 India, 1989.*

[68] Sherratt, J. A., Periodic Waves in Reaction Diffusion Models of Oscillatory Biological System, *Forma,* Vol.11 (1996), 61-80.

[69] Smith, G. D., Numerical Solution of Partial Differential Equations: Finite Difference Methods, *2nd edition, Oxford University Press, 1965.*

[70] Smith, I. M.; Farraday, R.V.and O'Conner, B. A., Rayleigh-Ritz and Galerkin finite elements for diffusion-convection problems,*Water Resources Research*, Vol.9, No.3(1973),593-606.

[71] Sun, M.; Tan, Y. and Chen, L. Dynamical behaviors of Brusselator system with impulsive input, dimensional diffusion-convection equation*, water resources research,* Vol.6 No.1(2007), 204-210.

[72] Sun, M.; Tan, Y. and Chen, L., Dynamical behaviors of the Brusselator system with impulsive input, *J.Math. Chem.*, Vol.44, No.3, (2008), 637-694.

[73] Turing, A. M.,"The chemical basis of morphogenesis" *Philos. Trans. R. Soc. London, Ser. B*, Vol.237, 37-72, 1952.

[74] Twizell, E. H.; Gumel, A. B. and Cao, Q., A second order scheme for the "Brusselator" reaction-diffusion system, *Journal of mathematical chemistry*, Vol.26 (1999), 297-316.

[75] Vacarezza, L. M.; Lombardi, J. L. and Chirife, J., Kinetics of Moisture Movement during Air Drying of Sugar Beet Root*, J. Food Sci.*, Vol.9(1974), 317-321.

[76] Walsh, J. B., *Finite element methods for parabolic stochastic PDE's, Potential analysis*, Vol. 23(2005), 1-43.

[77] Wazwaz, A.-M., "The numerical solution of fifth-order boundary value problems by the decomposition method," *Journal of Computational and Applied Mathematics,* vol. 136, no. 1-2 (2001), , pp. 259–270,

[78] Wiemar, J. R. and Boon, J. P., Non-linear reactions advected by a flow*, Physica A*, Vol. 224 (1996), pp. 207-215.

[79] Yang, L.; Zhabotinsky, A.M. and Epestein, I.R.,*Physical Review Letters*, Vol 92, No.19 (2004).

[80] You, Y., Global Dynamics of the Brusselator Equations, *Dynamics of PDE*, Vol.4, No.2 (2007), 167-196.

[81] Zaikin, A. N. and Zhabotinky, A. M., Concentration Wave Propagation in Two Dimensional Liquid Phase Self-oscillating System, *Nature, Lond.* , Vol.225 (1970), 535-537.

Printed by Books on Demand GmbH, Norderstedt / Germany